Principles of
A P P L I E D
MECHANICS

L. Eric Barefoot

Pearson
Custom
Publishing

Printed in the United States of America

10 9 8 7 6 5 4 3 2 1

This manuscript was supplied camera-ready by the author(s).

Please visit our website at www.pearsoncustom.com

ISBN 0–536–60418–5

BA 990837

PEARSON CUSTOM PUBLISHING
160 Gould Street/Needham Heights, MA 02494
A Pearson Education Company

Acknowledgments

This book would not have happened without the loving help of my wife, Mary. Her unwaivering belief in me, her support and patience kept my dream of completing this book alive. She has been my helpmate, my secretary, and my critic.

I wish to thank the many students and faculty at ECPI College of Technology for their input while this book was in draft form and those friends and family members who encouraged me.

I would also like to thank Fenton Priest of Priest Electronics (www.priestelectronics.com) for his assistance in obtaining permission to use the photographs from GC/Waldom Inc.

Never settle for less than your dreams.
If you work at it, some day they'll come true.

Strive for Excellence
Not Perfection

PURPOSE

This book is written to provide the beginning technician with basic information needed to repair or modify various types of electro-mechanical devices.

When making repairs on equipment, there are certain tools and techniques needed to perform those repairs. This book provides the technician with information on the use and function of many basic tools and mechanical parts found on such equipment. It also covers basic techniques used to make repairs.

This book was developed after conducting extensive employer research and personal experience. The employers expressed a need for the entry-level technician to be familiar with mechanical technology. The mechanical technology presented here has gained the approval of those employers. Once the student has mastered this technology, employers can expand the training on their specific equipment. It is therefore important for the technician to become familiar with the information contained in this book to quickly become productive in a technical environment.

Section I Wiring and Soldering covers different types of wiring, soldering tools, and soldering techniques. Section II is Hand Tools and Hardware and includes stock hardware, precision measuring devices, drilling, metal cutting tools, and taps and dies. Section III Mechanical Technology and includes mechanical terminology, mechanical ratios, electrical terminology, and mechanical switches.

TABLE OF CONTENTS

SECTION I   Wiring and Soldering

SECTION II   Hand Tools and Hardware

SECTION III Mechanical Technology

Answer Keys

Index

SECTION I – WIRING AND SOLDERING

CHAPTER 1 – BASIC WIRING

Introduction   When running or connecting electrical equipment of any type, consideration of hookup wires or connecting devices may cause confusion to the new technician. This Chapter on Basic Wiring will present the basic wires, wiring techniques, and connecting devices used in industry.

Objectives   After completing this chapter, you should be able to:

- Identify by gauge number a large diameter wire from a small diameter wire.

- Determine a gauge number without a wire gauge.

- Identify the different types of wires, their uses, and characteristics.

- Determine the purpose of terminal connectors.

- Identify basic types of terminal connections including how they are used.

- Determine the proper procedure to attach terminal connectors to wires.

- When given a set of adjustable wire strippers and a piece of wire, set the stripper so they will not damage the wire when used.

- When given a scale, tools, and wire, be able to measure and properly connect the wire to a screw type terminal.

- Identify the four basic types of wire splices and application of each.

- Be able to construct the four basic types of wire splices to a given set of specifications.

1-1   Standard Copper Wire Gauge

The purpose of a wire gauge is to determine the diameter of wires as a gauge number. The wire gauge, when used with the ***American Wire Gauge Chart,*** can be used to determine the current carrying capacity of wires.

Gauge Numbers – Wires are manufactured in sizes and are numbered according to a table called *American Wire Gauge* (AWG). The National Bureau of Standards sets the standards for wire manufacturing.

Wire Gauge – The wire gauge is a circular gauge with holes and slots in it (**Figure 1-1**). The wire to be measured is placed in the slot, **not the hole**. The gauge number is printed at the base of the hole. After the gauge has been determined, the table in **Figure 1-2** can be used to locate the information needed on specific gauge wire. In most cases, a technician will not be designing equipment, however, the technician should be familiar with wire gauges. If a piece of equipment is not running properly due to low current, it could be caused by incorrect wire size. The technician can check the wire gauge and the chart to determine if the proper gauge wire has been used before troubleshooting the equipment for low current problems.

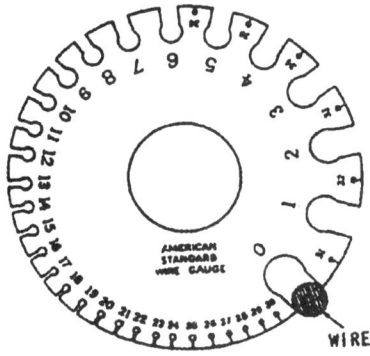

Figure 1-1 Wire Gauge

Determining Gauge Number Without A Wire Gauge – Another method of determining a wire's gauge is by using the American Wire Gauge Conversion Chart see **Figure 1-2**. The diameter of the wire is measured with a precision measuring device. Then, following the diameter in mils on the chart, the gauge can be determined.

Figure 1-2 is used to convert wire diameters to the proper gauge. All numbers are in circular mils that convert to ohms per 1000 foot of copper wire at 77°F. Gauge numbers range from 0-40. Remember, the larger the number, the smaller the wire.

GAUGE NO.	DIAM. IN MILS	CIRCULAR MIL AREA*	OHMS PER 1000 FT. COPPER WIRE @25°C	GAUGE NO.	DIAM. IN MILS	CIRCULAR MIL AREA*	OHMS PER 1000 FT. COPPER WIRE @25°C
1	289.3	83,690	.1239	21	28.46	810.1	13.05
2	257.6	66,370	.1563	22	25.35	642.4	16.46
3	229.4	52,640	.1970	23	22.57	509.5	20.76
4	204.3	41,740	.2485	24	20.10	404.0	26.17
5	181.9	33,100	.3133	25	17.90	320.4	33.00
6	162.0	26,250	.3951	26	15.94	254.1	41.62
7	144.3	20,820	.4982	27	14.20	201.5	52.48
8	128.5	16,510	.6282	28	12.64	159.8	66.17
9	114.4	13,090	.7921	29	11.26	126.7	83.44
10	101.9	10,380	.9989	30	10.03	100.5	105.2
11	90.7	8,234	1.26	31	8.928	79.70	132.7
12	80.8	6,530	1.588	32	7.950	63.21	167.3
13	72.0	5,178	2.003	33	7.080	50.13	211.0
14	64.1	4,107	2.525	34	6.305	39.75	266.0
15	57.1	3,257	3.184	35	5.615	31.52	335.0
16	50.8	2,583	4.016	36	5,000	25.00	423.0
17	45.3	2,048	5.064	37	4.453	19.83	533.4
18	40.3	1,624	6.385	38	3.965	15.72	672.6
19	35.9	1,288	8.051	39	3.531	12.47	848.1
20	32.0	1,022	10.15	40	3.145	9.88	1,069

Circular Mil - *The circular mil is the standard unit of measurement of a wire's cross-sectional area. This unit of measure is found in both American and English wire tables. A mil is one-thousandth of an inch or 0.001-inch. One circular mil is the cross-sectional area of a wire with a diameter of 1 mil.*

Figure 1-2 American Wire Gauge Conversion Chart

EXAMPLE FOR CONVERSION USING THE CHART IN FIGURE 1-2:

To find the area in circular mils of a piece of wire with a diameter .0201", first, convert the .0201" diameter in mils by moving the decimal 3 places to the right. This converts to 20.1 diameter in mils.

Next, square the 20.1 mils which equals 404.0 circular mils.

The circular mil number can now be converted to a gauge number by checking the chart in **Figure 1-1**; the gauge number is 24 gauge.

Checking the chart, a 24-gauge wire measures 26.17 ohms per 1000 feet of wire, which converts to .0261 ohms per foot of 24-gauge wire.

Summary

To know the resistance of wires is useful in determining the current capacity of wires. In the operation of equipment, sometimes the wires used to supply the power are not large enough. This can result in the malfunctioning of equipment due to low current. In addition, when the wires are too small, the high current can cause the insulation to melt, thus causing a short circuit or arcing. It is important for the technician to know the different gauges of wire and be able to determine their current capabilities using either the wire gauge or the *American Wire Gauge Chart*.

1-2   Types of Wire Conductors

A conductor is a physical piece of material that is used to connect components to form an electrical circuit. The circuit is used to perform a specific function for the operator. Without conductors, electrical circuits would not function.

Conductors come in a variety of physical sizes as well as combinations of those sizes. This section will present a few basic types of conductors, a picture of each, and an application for each.

Solid Wire – Solid wire is used in fixed wiring applications. A fixed application would be a machine or a piece of electronic equipment that requires a permanent internal wiring connection. In fixed applications the wires, when formed, retain that form, are not moved, and do not receive an excessive amount of vibration (see **Figure 1-3**).

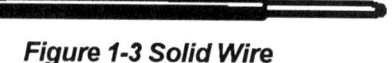

Figure 1-3 Solid Wire

- Solid wires have many applications from low voltage microvolts to high voltage power lines, with voltages in the thousands.

- Solid wire comes in all gauges and is insulated with many types of materials such as rubber, Teflon, thermoplastic, PVC, vinyl, etc.

Stranded Wire – Stranded wire is probably the most versatile of wires. It can be used where wires are moved or vibration is occurring. Stranded wire is made up of many thin strands of solid wire. Due to this type of construction it becomes more flexible. With normal usage, it will not break as quickly as solid wire of the same diameter. This type of wire is used on test leads for meters, extension cords, power cords, and many places where movement of wire will occur (see **Figure 1-4**).

Figure 1-4 Stranded Wire

- Stranded wire is coated with the same materials as solid wire. The voltage ratings are the same as solid wire. The only difference is that it consists of many small strands of wire, which gives it flexibility.

Wire Braid – This type of wire is used as a ground connection between sub units on machines. The wide braid is used to remove static electricity by providing a path to ground. Without a good ground path, electronic components could malfunction. This is especially prevalent on RF (Radio Frequency) equipment. The braid is also very flexible and is able to withstand extensive vibration. **Figures 1-5** and **1-6** show two different braid patterns. The patterns vary with manufacturer and perform the same function.

Figure 1-5 Wire Braid

Figure 1-6 Wire Braid

Coaxial Wire – This type of wire is also used for RF connections (Radio Receivers). The center conductor can be solid or stranded. In addition it can have several center conductors, depending on its application. The center conductor is insulated with various types of materials such as polyethylene or Teflon dielectric. The center material, the dielectric, is then coated with a copper shield. This shield is used to ground unwanted RF signals. **Figure 1-7** is a single conductor coax wire. **Figure 1-8** is a two-conductor coax wire. **Figure 1-9** is a multi-conductor with shield wire. In coax wire the number of conductors refers to the center wires, the shield is not counted as a conductor.

Figure 1-7 Single Coax

Figure 1-8 Double Coax

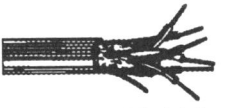

Figure 1-9 Multi Coax

In areas where the coaxial wire is to be in a fixed position, solid center conductors are used. Cable TV uses a solid center conductor. The cable is installed one time and is not moved. The cable connecting the VCR and the TV has a stranded-center conductor. This cable is moved many times and is difficult to break, whereas, if solid center conductors were used, breakage would be higher. Computers that are connected in a network use the stranded-center conductor cables.

Twin Lead – The twin lead hookup wire has many applications. It can be used for transmission line cables, antenna systems, TV, and FM multiplex (**Figure 1-10**). It is used for loud speaker connections and other low voltage applications such as alarm, signal, and telephone circuits (**Figure 1-11**). It is made at a fixed dimension. By keeping the wire at a fixed distance, interference from outside signals can be minimal.

Figure 1-10 TV Twin Lead   *Figure 1-11 Twin Lead*

Twin lead wire can have solid or stranded conductors depending upon its application. For fixed applications, the solid wire is used. When movement or vibration will occur, stranded wire is used.

Flat Cable (also known as ribbon cable) – The flat cable uses low voltage and is a small gauge wire (28 gauge). It is used in computers to connect the printed circuit boards together or to connect the disk drives to the printed circuit boards. It provides many connections at one time. **Figure 1-12** is a flat cable top view. **Figure 1-13** is an end view of a flat cable.

Figure 1-12 Flat Cable   *Figure 1-13 Flat Cable Side View*

Flat cable is normally made with stranded wire. Solid wire cables would not work inside the computers, as the dimensions in the computers are close and they would break. The cables must be twisted and formed to fit around many close areas, which requires flexibility.

Flat cables are not used exclusively in computers but are found in many pieces of electronic equipment or copying machines where many connections are needed at once for connecting controller boards to the main board.

Summary

Wires are used for a multitude of applications in electronic and electromechanical equipment. They come in all sizes, shapes, and construction. They all have one purpose: to provide a reliable electrical path for current to flow. The wires presented in this section represent only the basic types and their general applications. The wires presented and the application of each is not exclusive and they can be used in any manner deemed necessary by the designers of equipment.

It should be noted that wires break from improper usage. However, in most cases they can be repaired. Wires can be connected together with many types of devices. They can also be repaired with splices. Wires are essential in electronics and the technician should know how to connect and repair wires properly to provide a reliable electrical connection. Chapters 2 and 3 discuss wire repair.

1-3   Terminal Connectors, Tools, and Mounting Devices

Terminal connectors are devices that are used on the end of wires to make a connection on a terminal device. A terminal device is used as a junction or connecting point for circuits. The terminal connector provides a positive connection between the end of a wire and the terminal device. Terminal connectors are constructed in a variety of sizes and shapes which can be used in a variety of applications. There are basically two categories of terminal connectors – crimped which is insulated or soldered which is not insulated. Terminal connectors are designed to be crimped or soldered but not both.

Crimped type connectors have an insulator over the portion of the connector that covers the wire. It is used to protect the end of the wire as well as provide a strong connection. The terminal connector is then attached to a terminal block. The insulator refers to a material that will not conduct electricity.

The solder type terminal connector requires, as the name implies, soldering. This type does not have an insulator to cover the end of the wire. Therefore, when using this type of terminal connector, caution should be taken not to short the uninsulated portion to ground.

Terminal connectors have a color-coded nylon sleeve, and they are made of copper with tin plating to prevent corrosion (**Figure 1-14**). Some are made of tin-plated brass. The color of the sleeve denotes the size of the wire to be used with the terminal.

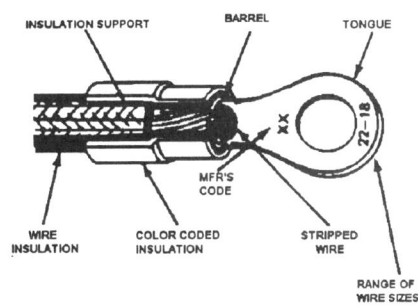

Figure 1-14 Terminal Connector

Terminal connectors also come in stud sizes (size of the screw). It is sometimes printed on the terminal connector depending on the manufacturer. Terminal connectors are grouped according to their application.

There are three basic groups: ring terminals, spade terminals, and quick disconnects. **Figure 1-14** shows a terminal connector using ring terminal. Ring terminals are used when the terminal lug is not normally removed thereby providing a permanent connection, **Figure 1-15** is an uninsulated ring terminal, and **Figure 1-16** is an insulated ring terminal.

Figure 1-15 Uninsulated Terminal Ring

Figure 1-16 Insulated Ring Terminal

Spade Terminals are used when the terminal lug may occasionally be removed. The spade means that the screw does not have to be removed to make the connection. **Figure 1-17** is an insulated spade terminal, and **Figure 1-18** is an uninsulated spade terminal.

Figure 1-17 Insulated Spade Terminal

Figure 1-18 Uninsulated Spade Terminal

Quick disconnects are used when the terminal lug is to be an inline connection. The term "inline" refers to a wire or several wires used to connect two pieces of a sub assembly. For example, a machine has a fan mounted on the inside cover. When service is required, the cover is removed. The wires connecting the fan and the chassis would have inline connectors. With inline terminal connectors, the cover could be completely removed to permit easy access to the machine. Quick disconnects require a mated set, a male and a female. They can also be insulated or uninsulated depending on the application. **Figure 1-19** is a male insulated quick disconnect, and **Figure 1-20** shows the female insulated quick disconnect.

Figure 1-19 Male Quick Disconnect

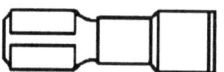

Figure 1-20 Female Quick Disconnect

Crimping Tools

A crimping tool is used to connect the terminal connectors to the wire. There are many types of crimping tools available. Each type will secure the terminal connector to the wire. The crimp performed by each tool is different. Some tools, like the one shown in **Figure 1-21**, require the operator to make two crimps – one on the wire and the other on the insulation. This type of crimp is considered weak due to lack of a positive control. It relies on the strength of the operator to apply the proper pressure. A precision crimping tool (**Figure 1-22**) will crimp both parts of the terminal connector at the same time. It also has a ratchet mechanism that requires the operator to crimp to a required pressure before it releases.

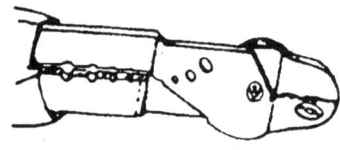

Figure 1-21 Single Crimp

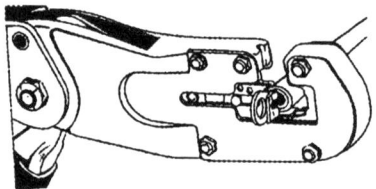

Figure 1-22 Dual Crimp

The technician must decide which tool is best for the job. If it becomes a matter of cost, the dual crimp tool is much more expensive than the single crimp tool.

Terminal Blocks

A terminal block is used as a junction point for many wires. The wires could come from subsystems within electronic or electromechanical equipment. This device is designed for many types of terminal connections. It can be for either mechanical or solder type terminal connections, and/or for single-row or double-row terminals. Terminal blocks can be used to connect 2 to 30 circuits and could be special ordered for as many as would be needed for a particular project. On any type of terminal block, the block size is dependent upon the voltage and current requirements of circuits. **Figure 1-23** shows a double-row, screw type terminal block.

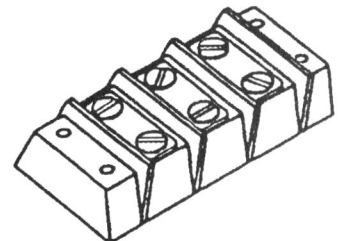

Figure 1-23 Terminal Block

Solder Type Terminal Strips – This terminal device is used for solder type connections. When used, it becomes a permanent connection point. Unlike the mechanical type of terminal block where the connections can be changed around more easily, the solder connection is fixed. This type of terminal strip can have a ground connection through the mounting hole or can be insulated from ground.

In **Figure 1-24** the terminal strip is the insulated type. This type does not have a provision for a wire to be connected to ground; thus it is referred to as an ungrounded type of terminal strip. **Figure 1-25** is of a grounded type of terminal strip. Note that one of the terminal connections would be connected to the mounting screw, thus it is said to be grounded (metal plate or common point on an electrical piece of equipment).

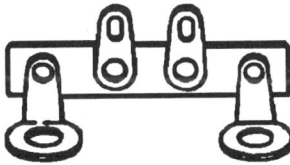

Figure 1-24 Ungrounded Terminal Strip

Figure 1-25 Grounded Terminal Strip

"D" Connectors – The "D" connector gets its name because it is shaped like the letter D, and it is an abbreviation for dataphone connector. The advantage of this connector is that it is used to connect as low as nine contacts to as high as fifty, or through special order, a greater number of contacts. They can be either a plug or socket type connector. The male is the plug and the female is the socket. The voltage rating varies with each type of connector. Many "D" connectors are rated up to 1250 volts AC and have current ratings up to 5 amps. **Figure 1-26** is a plug connector and **Figure 1-27** is a socket connector. Many are used on computers as connection cables.

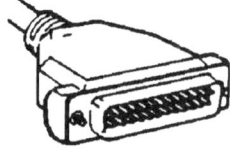

Figure 1-26 Plug Connector Male

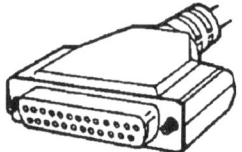

Figure 1-27 Socket Connector Female

Snap Splice – This type of terminal connector is used on high current lines. It is used to connect two lines together without soldering or crimping. It has a metal piece inside of it with two slits that, when pressed over the insulated wires, cut through the insulation and onto the copper wire to make the connection. It is then covered with its own plastic cover to prevent the wires from separating and also to prevent a short circuit.

The snap splice is used as a temporary connection. Extended use of the snap splice could result in corrosion between the metal parts resulting in an intermittent condition. The wires should be removed and soldered if a permanent connection is needed. **Figure 1-28** is a snap splice. Note the metal in the center of the splice that is used to connect the wires.

Figure 1-28 Snap Splice

Wire Nut – The wire nut is used to connect many wires together without the necessity of soldering. It is constructed of an insulated material with a metal spring with groves placed inside. When placed over the wires and tightened, the grooved metal insert twists the wires tighter and prevents them from falling out of the holder; thus, it is a wire nut (**Figure 1-29**). This device is normally used on high current wires where soldering to a terminal connector or using a terminal block and lugs is not needed and would increase the cost of the project.

Figure 1-29 Wire Nut

Wire nuts are used on AC lines and some DC lines. With wire nuts, disconnecting and reconnecting of power lines is easy.

Butt Splices – A butt splice is a device used to connect two wires. This device requires a crimp and is not soldered. Because it is crimped, the current requirements are higher than soldered connections.

Figure 1-30 is a butt splice. They are usually color coded for different wire sizes. Internal construction is a metal sleeve inside an insulated sleeve or barrel. The wires are butted together in the metal sleeve, then crimped. The insulation covers the bare wires preventing shorts. Note: The picture above the butt splice shows how the wires are put into the butt connector and crimped.

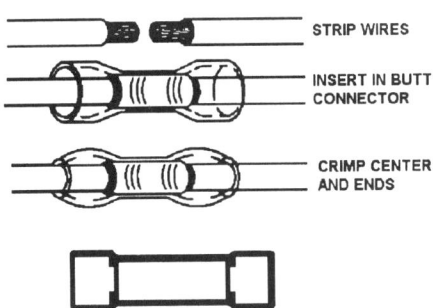

STRIP WIRES

INSERT IN BUTT CONNECTOR

CRIMP CENTER AND ENDS

Figure 1-30 Butt Splices

It is a quick method for reconnecting broken wires, however, it should be noted that when reconnecting lines, the butt splice takes up more physical space than a soldered splice. In some machines or electronic equipment, space is at a premium and using the butt splice should be used with caution. Butt splices are sometimes called barrel connectors. In some places it is not advisable to use them as corrosion can take place if they are not properly sealed from the elements.

Jacks and Plugs – Jacks and plugs refer to physical connectors that are used to temporarily connect electrical points. Plugs are the female connectors and jacks are the male connectors. When referencing this type of connector, the term jack is used. The plug is assumed to be part of the jack. There are hundreds of jacks available. Each manufacturer designs their jacks to be unique. In some cases this uniqueness is necessary to prevent unauthorized connections. There are many universal jacks. Two of the most commonly used are banana jacks and tip jacks.

When choosing jacks, it is advisable to select one that can be repaired. Many jacks have molded plug ends and, although they are much safer to use, they cannot be repaired. When used extensively, breakage can occur.

Banana Jack – One type of jack is called the banana jack. This type can be repaired. It has either a screw or solder type connection. The jack has a spreader end which, when inserted into the plug, acts like a spring and remains tight. The jack, after extended usage, becomes compressed. It can be repaired by simply inserting a small screwdriver blade between the compressed portion and the center shaft and reshaping the compressed portion. Thus the banana jack is reusable.

The banana jack is used on electronic test equipment. The design offers few maintenance problems and provides a good contact for electronic connections. **Figure 1-31** is a banana plug and **Figure 1-32** in the mating banana jack.

Figure 1-31 Banana Plug   *Figure 1-32 Banana Jack*

Tip Jack – The tip jack is another versatile connector. It must be connected to the wire with solder. This permits ease of repair. The tip jack is a solid pin that does not require maintenance. The tip plug does wear from usage. Some types of plugs can be repaired and other types cannot.

The advantage to this type is its versatility and low maintenance. It is primarily used on electronic test equipment that requires connecting and disconnecting many times. **Figure 1-33** is the tip plug and **Figure 1-34** is the mating tip jack.

Figure 1-33 Tip Plug   *Figure 1-34 Tip Jack*

Summary

The term terminal connector means the end of the wire or the end of the connection. There are many devices that provide reliable electrical connections. With the appropriate terminal block, they provide uniformity of construction. There are a variety of terminal connectors that can be used to provide reliable electrical paths between electronic components. The technician should know the basic types discussed in this chapter and should make it a point to become familiar with the many others available on the market.

1-4   Exercises

Exercise A – Wire Stripping

The purpose of this exercise is to familiarize the technician with the proper method of setting a standard adjustable pair of wire strippers. Improper setting of the wire strippers can result in nicking the copper wire. A nick, depending on the depth, can result in the wire's current capabilities being changed. In electronic terminology, the wire's resistance (opposition to current flow) will increase. This results in less current flow at that point. In addition, the current requirements of the wire could be high and, at the point of the nick, could cause the wires to get hot enough to melt the insulation that can cause a fire. The nick in the copper wire can also cause a weakening at that point which can result in premature breakage.

Adjustment Procedure – The wire stripper used in this exercise is the adjustable type. It has a slot in the side with an adjustment screw that is used as a stop for the handles. This prevents the wire stripper from closing too far and cutting or nicking the copper wire. A slotted screwdriver is used to make the adjustment. **Figure 1-35** shows the location of the screw. The following steps are used to properly set the strippers.

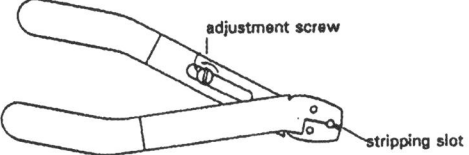

1.   Cut off enough wire to use for practice. With a knife, remove approximately ¼" of insulation from the practice wire. This is to expose the copper wire that will be used to set the strippers.

2.   With a slotted screwdriver, loosen the adjustment screw and open the stripping slot enough to insert the wire into the slot. Carefully close the stripping slot until the cutting slots just touch the wire. The wire should be able to move within the stripping slot (**Figure 1-36**) just slightly scraping the sides of the wire. Now tighten the adjustment screw.

Figure 1-36 Setting the Adjustment

3.   Test the strippers with a new piece of wire.

    a.   Make a cut ¼" cut from the end of the wire by closing the stripping slots until they are at its stop, then release the stripping slot.

    b.   Make the second cut by ROTATING the wire half the distance from the first cut. Release the stripping slot and remove the wire from the strippers.

This procedure is done to ensure a maximum allowable cut on the wire. Note: The wire is round and the cutting slots are in a diamond shape. This means that the cutters will not completely cut the insulation on the two cuts.

4.   To check the wire strippers to ensure that they are not too tight or too loose, make a second cut ¼" from the last cut. Remove the wire from the strippers. Grasping the wire, twist the second cut and pull both pieces at once (**Figure 1-37** below).

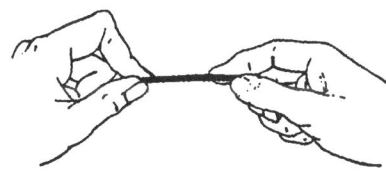

5.   Check the wire for nicks. If a nick is present, repeat step 2 and check again. When the strippers are set properly, the insulation will be removed easily and there will be no visible nicks on the wire.

Figure 1-38 shows a wire properly stripped of its insulation. **Figure 1-39** shows a wire that has been improperly stripped (notice the nick in **Figure 1-39**).

Figure 1-38 Correctly Adjusted   *Figure 1-39 Incorrectly Adjusted*

Exercise A Summary

It is important that the technician practice this exercise to become proficient at setting the wire strippers properly. The technician will be making repairs on broken cables or modifying equipment and there may not be enough wire to practice with or the space to work in is tight, which requires that the repair be made properly the first time. Many technicians do not take the time to practice and when time requires proficiency, quality is lost.

Exercise B – Wiring Termination

This exercise is designed to familiarize the technician with the measuring scale, converting fractions or decimals to fractional measurement, and pre-determining wire lengths using scale measurements. In addition, the technician will practice using those measurements to pre-form wires and then properly connect the wires to terminals without using terminal connectors.

There are times when terminal connectors are not needed and, if used, could increase the cost of repairs or modifications. Therefore, it is important that the technician practice connecting wires to terminals without terminal connectors, so that when tasked, the connections can be made properly and efficiently.

For this exercise, a terminal block with four circuit connections was chosen with a number six screw (**Figure 1-40**). Other terminal blocks will work as long as the technician recomputes the different screw size and follows the same procedure using the nominal lengths given in this exercise.

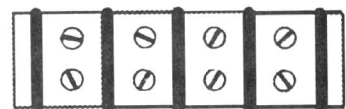

Figure 1-40 Terminal Block

If terminal blocks with larger screws are used other than the numbers recommended, use larger gauge wire for this practice. Since this is a practice session, the wire to be used can be any gauge between 18 and 24. This will provide good material for practice.

When tightening the screws on the terminal block, the screw driver blade should fit the screw slot with minimum side-to-side movement. This will prevent the screwdriver from slipping out of the slot and causing injury to the operator or damaging the terminal blocks.

Scales – Scales are measuring devices that divide the inch measurement into fractions. Standard scales use 64th, 32nd, 16th, and 8th divisions. **Figure 1-41** is a 1/16 scale. Each scale number represents the number of divisions in one inch.

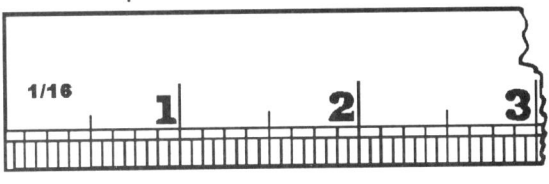

Figure 1-41   1/16" Scale

Example:
> Using the 16th division, there are 16 even divisions in one inch.
> Using the 32nd division, there are 32 even divisions in one inch.

> When making measurements, it is advisable to choose a division that will be easily read. In determining measurements, allow for a read error of plus or minus one division. This procedure is for wiring techniques and does not pertain to precision measurements.

Converting Fractions and Decimals

> Convert a FRACTIONAL part of an inch to a DECIMAL.

Example:
> Convert 1/16 of an inch to a decimal.

> Always divide the top number by the bottom number, to change a fraction to a decimal.

> Divide the 1 by 16 which equals .0625.

> Convert a DECIMAL to a FRACTIONAL part of an inch.

Example:

Convert .8125 to a fraction.

When converting, determine the number of divisions desired and use the decimal equivalent for that fraction.

1/16 of an inch is .0625.

Divide the .8125 by .0625 and the answer is 13. Therefore .8125 equals 13/16 of an inch.

To convert to 32nd.

Convert 1/32 to a decimal, which equals .03125. Then divide .8125 by .03125 and it becomes 26/32 and it reduces to 13/16.

Pre-Determining A Length of Wire

Figure 1-42 is the wire form needed to connect to the terminal block. Note: it must be formed around the screw. To prepare a wire to be connected to the terminal block, all measurements must be added.

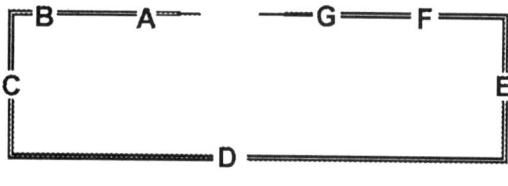

Figure 1-42 Wire Form

NOTE: The dimensions for the wire were determined using the four circuit terminal block with a number six screw.

- The amount of bare copper wire needed to go around the screw is determined by the diameter of the screw times the constant 3.1416.
- The diameter of a number 6 screw has an average measurement of .138.
- Multiply .138 X 3.1416 = .4335.
- Divide .4335 by .0625 which equals 6.936, round up to a 7 and it becomes 7/16.
- Add the lip of the screw head, that is 1/16 inch.
- Add 1/32 for bare copper to be visible.
- Add all measurements, 7/16 + 1/16 + 1/32 = 17/32 and add 1/32 for read error.
- The total length of insulation to be removed is 18/32 that reduces to 9/16 of an inch. This is dimension A (this dimension is the same for G).
    B - C - E - F

- "D" dimension is determined by measuring the distance between the screw heads on the terminal block (**Figure 1-43**).

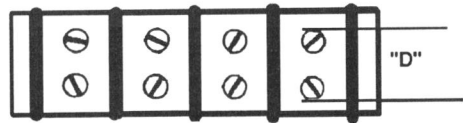

Figure 1-43 Terminal Block

- Add B and F dimensions.
- Add D, the dimension should be 10/16". If not 10/16", use your measured dimensions as variations may occur between terminal blocks used.
- When adding ½" for B and ½" for F, and if D equal 10/16", then the **total** for B, F, and D should be 1 and 10/16" inches.

Figure 1-44 is the wire form in a straight line with all pre-determined dimensions. To determine the total length needed, add A through G. The total length is 5.75", which converts to 5-3/4".

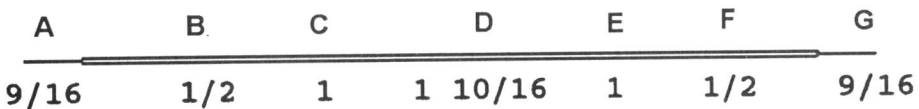

A	B	C	D	E	F	G
9/16	1/2	1	1 10/16	1	1/2	9/16

Figure 1-44 Wire Form in a Straight Line

Preparing and Attaching Wires

1. Cut a wire 5 ¾" long.

2. Remove 9/16" insulation from each end.

3. Use a pair of long nose pliers and bend the bare copper on each end in a loop to fit around the (**Figure 1-45**). The loop should be formed in the direction that the screw will be turning. To form the loop in the other direction will force the wire away from the screw and cause a poor connection.

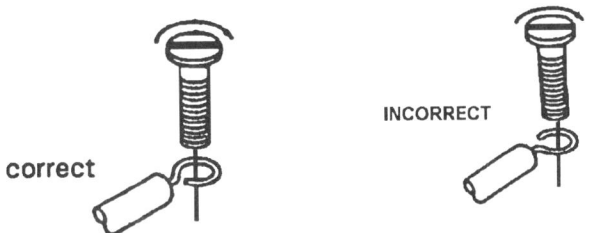

Figure 1-45 Loop Around Screw

4. From the insulation, measure ½" and bend the wire 90°.

5. From the first bend, measure 1 inch and bend the wire a second 90°.

6. Next measure 1-10/16" and make the next bend.

7. Repeat step 5 then 4 on the opposite of the end of the wire.

8.   Remove the two end screws from the terminal block (**Figure 1-46**). Place the screws through the two pre-formed end loops and replace screws in the terminal board. If all measurements have been followed, the wire form should fit properly. If it does not, recheck measurements.

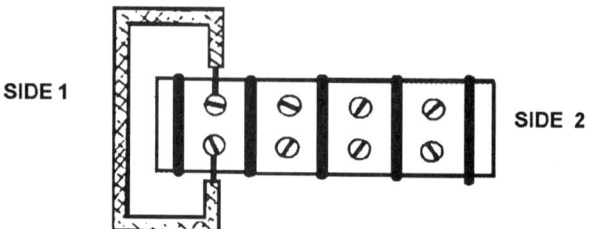

Figure 1-46 Terminal Block with One Wire

9.   Repeat the same procedure to pre-form a wire for the second side (**Figure 1-47**).

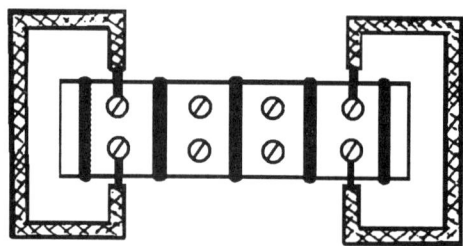

Figure 1-47 Terminal Block with Two Wires

10.   In **Figure 1-48**, a third wire has been added.

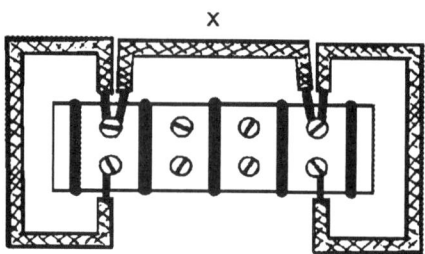

Figure 1-48 Terminal Block with Three Wires

11.   Measure the distance "X" between the end wires as shown in **Figure 1-49**.

Add ½" to extend from the screw and the 9/16" for the screw since there are two connections. Multiply those measurements by 2 and add distance "X" to obtain a total.

Cut a wire to the total length and remove the insulation from each end.

Pre-form the wire, then attach to the terminal block. Be sure that the loop is formed in the direction of the screw's rotation.

12. Make a second wire the same length as in step 10. Prepare wire and connect it to side "Y".

13. Upon completion of this step, the final wiring diagram should look similar to **Figure 1-49**.

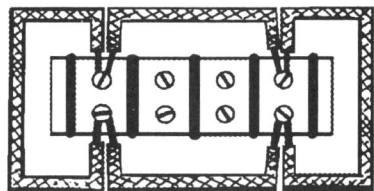

Figure 1-49 Finished Product

Exercise B Summary

This was a practice exercise that demonstrated the use of a scale and a basic wiring procedure that will enable the technician to properly connect a wire to a terminal block or any other terminal connection using a screw type connection. It also provides technicians with preset wiring instructions. Many times reading and following instructions is not done and can result in the device not being according to specifications. LEARN TO FOLLOW INSTRUCTIONS.

Exercise C – Wire Splicing

There are times when cables or individual wires supplying power to your equipment becomes damaged or a modification may be needed to update equipment. To repair the cable or wires, splicing may be necessary as well as cost effective over other mechanical devices needed to perform the same job. In repairing cables or performing modifications on equipment, there are several methods that can be used to connect wires. Some of the mechanical devices are wire nuts, butt splices, barrel crimps, or wire splicing. All use a mechanical connection of some sort. The wire splice is the most economical of the group and it is the most positive of the group. That is, it has both a mechanical connection and it is soldered. It can be used for low voltage, low current, or high voltage, high current wires. It is much more versatile than the mechanical device. It saves space, and it can be made in such a manner that it appears similar to the original wire.

There are four basic splices with which the technician should become familiar. They are the *PIGTAIL JOINT, WESTERN UNION JOINT, KNOTTED TAP JOINT*, and the *FIXTURE JOINT*.

- Each splice has a specific use.
- Each requires practice time to master.
- Practice splices can be soldered.
- They will provide the technician with items on which to practice soldering techniques.

The following basic splices should be practiced until they can be made properly each time. The **recommended** wire to practice with is 24 to 18 gauge, solid, and a single conductor copper wire.

The quality of connections (splices) depends upon the individual. Some can master the mechanics of the connection quickly, and others may take longer. It is recommended that a dozen of each be made, not only for practice but to provide the technician with a good supply of practice connections for soldering techniques. Improperly constructed connections should not be discarded, as they will also provide items with which to practice soldering techniques in Chapter 3.

1.   **Pigtail Joint** – This connection is the most common of all connections. It is the easiest to make, and it can be used on any gauge wire. It can be soldered and re-insulated, or it can be used with a wire nut unsoldered. The wire nut, when placed on the end of the wire, is screwed onto the wire causing the pigtail to become tighter. The disadvantage is that, after construction, it is bulky and difficult to form into any cable situation requiring the wires to be flat.

This connection is normally used where there is plenty of space available and the connection will be inside a covered housing.

The pigtail is usually made with solid wires, but it can be made with stranded wires. It is recommended that if stranded wires are used that the end of the stranded wires be tinned (coated with solder) to prevent fraying. If tinning is not practical, care should be taken not to break the strands. Care should be taken with loose strands for they can touch other wires that could result in short circuits.

Construction:

This exercise requires solid wires and any gauge wire between 18 and 24. For this exercise, 24-gauge wire is used.

The proper way to construct a pigtail is to first remove approximately ¾ of an inch of insulation from each wire to be spliced.

Form an "X" (**Figure 1-50**) with both wires and twist both wires together at the same time. This will form a tight connection. DO NOT twist one wire around the other in the form of a candy cane. This will cause the wires to come apart when moved.

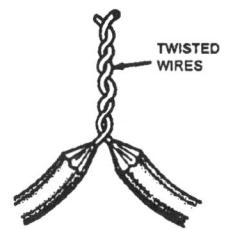

Figure 1-50 Pigtail Splice Forming "X"   *Figure 1-51 Complete Pigtail*

After twisting both wires to the end (**Figure 1-51**), measure from the insulation to the end approximately ½" and cut the rest from the end. As wire sizes vary, adjust the cut to accommodate the situation.

After completion, re-insulate the wires with electrical tape or use a wire nut. If tape is used, be careful to choose the correct type of electrical tape for the electrical characteristics for your wires. Not all electrical tape is the same. Each type is voltage rated. Check with suppliers for voltage ratings on tapes.

2.   **Western Union Joint** – This splice is used to repair cables, or to make inline splices where the splice must follow the diameter of the wire, and when re-insulated can be made to look like the cable system. When re-insulating this splice, electrical tape can be used but it will be bulky and not allow the splice to become contoured to the wires. The best method to re-insulate this splice is with shrink tubing. Shrink tubing comes in a variety of colors and sizes. Shrink tubing, when heated, forms itself to the wire and becomes a tight connection. Since it forms to the wire, it will look more like the wire and will not be bulky. It can also fit better in the original cable. Spaghetti tubing can be used, but caution should be observed as it can slide off the connection if the wires are moved, or there is excessive vibration of the cables or wires.

Construction:

To construct the western union splice, perform the following steps:

Note: The following measurements are for 24-gauge solid wire. When using different gauge wire, measurements must be adjusted. The 1/32 rule is based on the diameter of 24-gauge wire and is used as a guide. The wire diameter is used to determine the distance from the end of the splice to the beginning of the insulation. It is equivalent to the diameter of the wire.

Example: The 24-gauge wire is 1/32" diameter; therefore, the distance from the end of the wrap and the beginning of the insulation should be 1/32". If 18-gauge wire were used, then the same distance would be 3/64".

Note that the measurements presented are given to act **ONLY AS A GUIDE** and **are not considered absolute**.

Figures 1-52 through **1-55** are to be used as a guide to construct a high quality connection. Each step is to act as a guide.

In **Figure 1-52**, the "X" being shown has had 1" of insulation removed from each wire. The "X" is started at ¼" from the insulation.

In **Figure 1-53** the "X" is twisted three turns leaving the two ends of each of the wires extending away from the center.

In **Figure 1-54**, the two wires are wrapped around the straight wires in a tight manner a total of four turns plus or minus one turn.

Figure 1-55 shows the completed splice. Note that the ends of the wires are formed completely around the wire with no tabs visible (the ends of the wires that are not completely formed to the round wire are called tabs). If the splice is constructed properly, the distance from the end of the splice and the

beginning of the insulation should be close to 1/32". Save wires for soldering practice in Chapter 3.

3.   **Knotted Tap Joint** – This connection is used for connecting to an existing line, thus the name TAP joint. The knotted tap gets its name from the actual construction of the joint. The first wrap made is a knot. This connection can be made with a snap splice rather than by the knotted tap splice.

The disadvantage of the mechanical snap splice (covered in 1-3) is that it is primarily used for high current circuits. This type of mechanical connection requires the connecting blade to cut the insulation of each wire as it is closed. The connection is made through the thin metal connecting blade. It is convenient but limited in applications. The knotted splice is more versatile and can be used on all sizes of wires with varying current requirements.

Upon completion of the splice re-insulation is achieved with electrical tape. If the splice is being connected to a preexisting wire, shrink tubing is not possible; however, shrink electrical tape is available, although it costs more than standard electrical tape.

Construction:

To construct this splice, 24-gauge wire is being used and all dimensions are based upon this gauge.

The first step is to remove ¼" of insulation from the main line (line being connected to). This is done by making two cuts with the wire stripper's ¼" apart.

Then, using a knife, cut between the two wire stripper cuts, removing the insulation. NOTE: Care should be used not to nick the wire.

Next, remove 1" of insulation from the branch wire that is to be attached to the main line.

Figure 1-56 shows the first wrap. The insulation should be 1/32" away from the beginning of the first bend on both the branch wire and the main wire. The bend goes from the back side to the front side of the branch as shown in **Figure 1-57**.

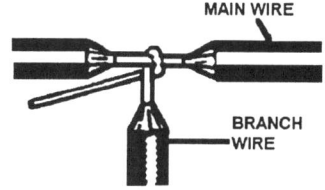

Figure 1-56 First Wrap

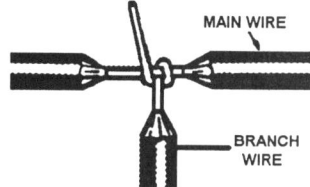

Figure 1-54 Second Bend

The branch wire then goes over the main wire as shown in **Figure 1-57**. In the next step (**Figure 1-58**), the branch wire is bent or wrapped around the main wire thus the knot is formed. The final step (**Figure 1-59**) shows the branch wire wrapped around the main wire five times. The number of wraps should be five plus or minus one turn or wrap. When completed the wrap should end as close to 1/32" as possible. Be sure to fold the end of the wire tightly around the main wire to prevent the end from protruding through any insulating materials.

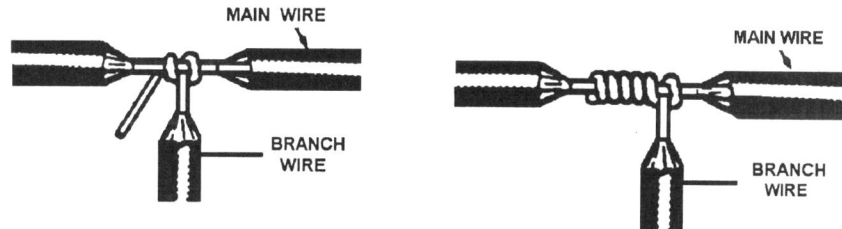

Figure 1-58 Knot Formed   *Figure 1-59 Final Wraps*

This completes the construction of the knotted tap joint. As mentioned earlier, construct as many as possible to achieve a properly constructed joint.

Note: Save all connections, good and bad, for soldering practice. To learn proper soldering techniques, many connections need to be made for practice. A good quantity would be around 12 to 20 connections.

4. **Fixture Joint** – This connection is used when two different gauge wires are to be connected. It is also used in place of the pigtail joint. There are times when the pigtail cannot be used due to one wire being larger in diameter than the other wire. The different sizes prevent the two wires from being properly twisted together.

    Construction:

    NOTE: For practice, use two wires of the same gauge. This exercise is based on using 24-gauge wire.

    Figure 1-60 shows a branch wire (the larger wire) and the fixture wire (the smaller wire). Since the fixture wire is smaller, it is to be wrapped around the branch wire many times. Therefore, the fixture wire should have the most insulation removed. When using other size wires, the ratio of insulation to be removed is approximately ½ the exposed length of the fixture wire.

    Example: If the fixture bare wire is 1", then remove ½" of insulation from the branch wire.

    Figures 1-60 to **1-61** show only a slight difference in the diameters of both wires. Therefore the wires being used for this exercise will both be 24-gauge wire.

    First remove 1-1/2" of insulation from the fixture wire, and then remove ¾" from the branch wire. (**Figure 1-60**). Be sure to maintain 1/32" bare copper from the end of the insulation to the beginning of the first wrap.

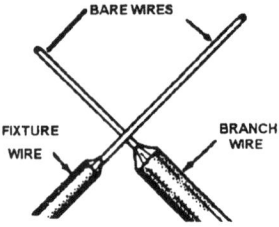

    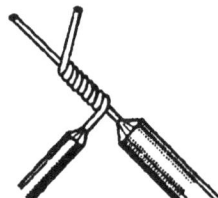

    Wrap the fixture wire around the branch wire tightly at least eight turns (**Figure 1-61**).

    Fold the branch wire over the fixture wire and wrap tightly (**Figure 1-62**) forming a hook.

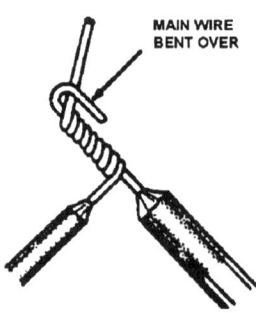

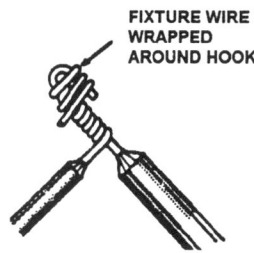

FIXTURE WIRE WRAPPED AROUND HOOK

Wrap the remaining fixture wire around the hook (**Figure 1-63**) and be sure to fold the end snugly.

This completes the construction of the fixture joint. This joint should be practiced until it can be made with ease. Remember to save all wires made for soldering practice.

Exercise C Summary

This completes the wire splicing exercise. The quantity and quality of the splices depends on the amount of time and care used to make them. The technician should be able to construct any of the basic splices with a minimum of difficulty with various size wires. The splices are useful to know since most splices made by untrained technicians are constructed improperly, which results in the splice coming apart at a later time. The splices should be saved for soldering practice (Chapter 3). Improperly made splices should **not** be discarded as they can also be used for soldering practice.

SECTION I – WIRING AND SOLDERING

CHAPTER 2 – SOLDERING TOOLS & MATERIALS

Introduction Soldering tools are used for soldering wire splices, replacing broken connections, and repairing printed circuit boards. The service technician is the one who will be making repairs, usually at the customer's place of business. There are times when the specialty tools are not available to make emergency repairs. However, there are devices that can be used in place of these tools. This Chapter points out those devices and makes reference to them. The soldering tools are designed for making professional looking repairs, as well as being used for other repairs such as construction of new products. It's important that the repairs look as close to the original as possible. If repairs are made using the proper tools and procedures, then the unit will function more reliably.

Soldering materials are materials used to make repairs to broken equipment or cables. They are used to make the connection like new. There are many materials manufactured for that purpose. The choice is determined by the design requirements of the manufacturer. If the manufacturer does not specify the type of materials to be used then it is up to the technician to choose the appropriate materials. In this section basic materials are presented.

Objectives After completing this chapter, you should be able to:

- Identify the basic types of soldering units and characteristics of each.

- Determine the proper soldering unit to use when soldering.

- Identify basic tools used to make repairs on broken equipment.

2-1   Soldering Tools

Soldering Irons

There are many types of soldering irons on the market. Some are portable, requiring a battery to heat the element, some use butane to heat the element, and others use AC power to heat the element. **Figure 2-1** shows all three basic types. The AC type is the most versatile. It can supply constant heat and can control the temperature of the tip more accurately than the portable irons. Each type is designed to get hot enough to melt solder. Solder melting range is from **361°** to **576°** Fahrenheit.

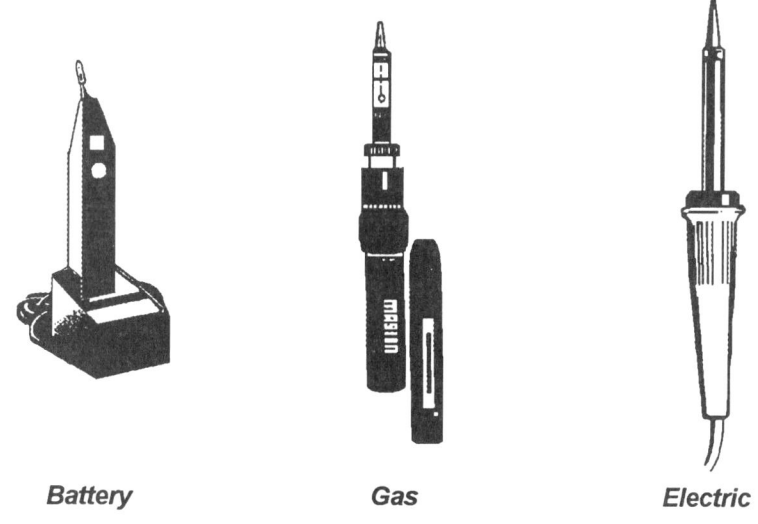

Battery   **Gas**   **Electric**

Figure 2-1 Three Types of Soldering Irons

The AC soldering iron element, that part which heats the tip, varies in wattage, just like light bulbs. They can vary from low wattage to high wattage depending on the type of soldering to be done. Some soldering irons have changeable elements that can vary in wattage. The correct one to use depends on the skill of the technician and the type of repair. It is recommended that for printed circuit board repairs, a 15W-35W iron be used. A 25W iron is the safest on printed circuit boards. For repairs such as cables and connectors, a higher wattage soldering iron may be needed depending on the size of the wire.

Note: The irons discussed are for servicing at the job site and for making repairs, not for high quality bench work.

Irons that are used for bench level repairs should be more specialized, with temperature controls for the tip and special de-soldering units to remove solder. This type uses a vacuum pump connected through the hollow tip. This type of unit is more expensive as well as much larger.

When soldering, the tip of the iron transfers the heat from the element to the part being soldered. The size and type of the tip is another important element to consider before soldering. Choose one that will fit the part to be soldered. If the tip is too wide, it could solder or heat other parts next to the one being soldered. If it is too narrow, then it may not transfer the heat properly and cause a defective connection.

Pencil Soldering Iron

This is the main tool used to heat the surface area to be soldered (**Figure 2-2**). A soldering iron from 15W-35W is good for electronic circuits. It should have a removable element and tip. This will give more versatility to the soldering iron. A 25W soldering iron is recommended for soldering electrical circuits on printed circuit boards. A soldering iron of 100 watts or greater is used for soldering large electrical connections. Some have an adjustable power source that is used for controlling the temperature of the soldering iron tip. A soldering iron stand is for supporting a hot soldering iron when not in use.

Figure 2-2 Pencil Soldering Iron

A soldering vise is used for clamping and holding a printed circuit board or other components during soldering or other repair operations.

Resistance Soldering Unit

The resistance soldering unit uses two electrodes that contact the component to be soldered. Heat is created through current flow (similar to arc welding) (**Figure 2-3**). It is used primarily for cable connector construction and is not recommended for printed circuit boards.

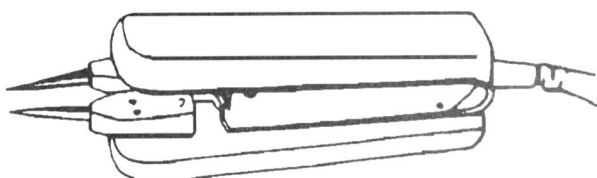

Figure 2-3 Resistance Soldering Unit

Soldering Guns

Soldering guns work on the same principle as the pencil tip soldering units, except that they normally use an AC transformer to heat the tip with high electric current. When using a transformer to produce heat, an electromagnetic field is formed. This magnetic field produces voltage. The voltage can damage components and integrated circuits on printed circuit boards. They are NOT recommended for printed circuit boards. They are recommended for use on high current connections.

2-2   Soldering Aids

One type of soldering aid is a long narrow tool with thin long blades for picking, moving, or prodding components where room is at a premium. The ends are made of aluminum. One end has a slot for bending component leads to form a proper fit (**Figure 2-4**). The other end has a thin blade used for picking at components or circuits.

Figure 2-4 Soldering Aid

Heat Sink

Another type of soldering aid is the heat sink. **Figure 2-5** is a spring actuated tool made of aluminum and used to clip on components or leads to help dissipate excess heat that could cause damage to components.

Figure 2-5 Heat Sink Soldering Aids

2-3   Wire Strippers

There are many types of wire strippers available. Their primary function is to remove insulation from wire without nicking the copper wire or damaging stranded wire by cutting strands. When selecting wire strippers for use, the application will determine the type and quality to be selected. Not all wire strippers are the same. Many of the inexpensive wire strippers are made of poor quality steel, and the jaws will bend easily. Good quality wire strippers will not bend as easily.

Mechanical Wire Strippers

Mechanical wire strippers are easy to use. They are an adjustable and an inexpensive tool. They look like a pair of scissors but have a cutout on both jaws for placement over the wire to remove insulation from conductors (**Figure 2-6**). The jaws are diamond shaped requiring the operator to make two cuts before removing the insulation.

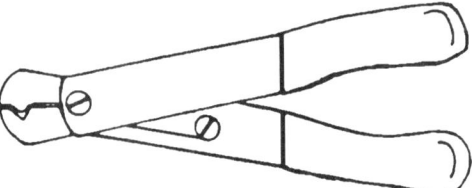

Figure 2-6 Mechanical Wire Strippers

Automatic Wire Strippers

The name, automatic wire strippers, refers to the mechanical action of the strippers. This type is also rated on quality. There are inexpensive automatic strippers (**Figure 2-7**). Their cutting jaws are "V" grooved and rely on the mechanical action of the strippers to aid in insulation removal. These are fairly inexpensive, but generally efficient. On a good pair of automatic wire strippers, the cutting jaws are made to fit the exact size of the wire. They are designed so that they can cut through the insulation without damaging the copper wire. The jaws are machined for precision. The jaws are changeable to accommodate different size wires and insulation. There are many different types of insulation including rubber, PVC, vinyl, and Teflon to cite a few. The automatic strippers are used primarily in cable work, where many wires are to be connected at one time. There are other types of automatic wire strippers such as those used for coax cable. This type will remove the exact amount of insulation without damaging the coax line. There are also specialty wire strippers.

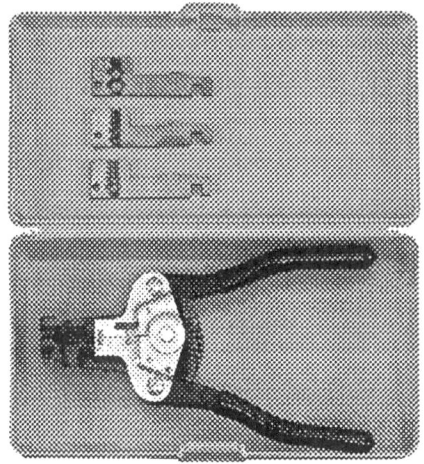

Figure 2-7 Automatic Wire Strippers
(Reprinted by Permission of GC/Waldom Inc.)

Thermal Wire Strippers

Thermal wire strippers are for removing insulation from the wire by melting the insulation material (**Figure 2-8**). This method prevents damage to the solid copper and will not nick stranded wires. Wire coated with polyvinyl chloride insulation will melt. The disadvantage of this tool is that it cannot be used on insulation materials that will not melt, such as glass braid, PVC, or Teflon insulated wires. Other disadvantages include the limited amount of insulation materials that can be removed by heat and, without a special attachment, can only be used on single stranded wire. This type of device is limited to production areas and is not a tool that can be carried to the job site.

Figure 2-8 Thermal Wire Strippers

2-4   Pliers

There are many types, sizes, and manufacturers of pliers. Pliers range from inexpensive to very expensive. Some are made of surgical steel and others are made of high quality steel. Each type is designed for a specific purpose.

Pliers were designed as an extension of your fingers. They are designed to cut and shape wire and to fit into a variety of areas. Without pliers, wires cannot be formed properly which could result in their interfering with each other, mechanically and electrically. They also come with insulated handles to prevent the operator from being shocked accidentally. The insulated handle also provides comfort to the operator.

If pliers are used properly, they will last many years. The quality of the pliers depends on the type of steel used in making the pliers. A good quality set of pliers can be very expensive, but will last and hold together longer.

Long Needle Nose Pliers

Long needle nose pliers have jaws that are very thin and long. Needle nose pliers can have short thin jaws that are used to position or bend component leads in tight places (**Figure 2-9**). Since the blades are thin, only small diameter wire should be formed with these pliers. If they are used on too large a diameter of wire, the pliers can be damaged. To be classified as needle nose pliers, they must be thin but the length is not a deciding factor.

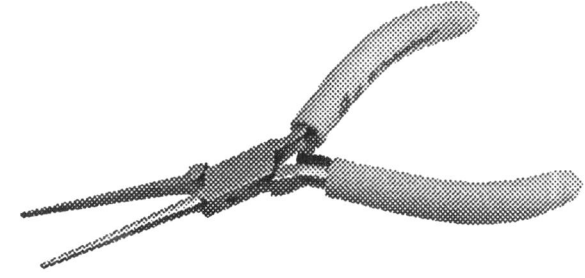

Figure 2-9 Needle Nose Pliers
(Reprinted by Permission of GC/Waldom Inc.)

Long Chain Nose Pliers

Long chain nose pliers have long tapered jaws, which are used to position and or bend larger diameter wire or component leads (**Figure 2-10**). The jaws are larger and thicker than the needle nose pliers. They are designed more for heavier duty wiring than the long needle nose pliers. They are used for making mechanical connections prior to soldering. They may also be used as a heat sink during soldering. The points of this type of pliers vary in thickness. Some are heavier on the end and others are very thin. Caution should be exercised when using this type of pliers. Be sure to use the proper type of pliers for the job. The tips can be damaged if improperly used. They are very versatile for making repairs.

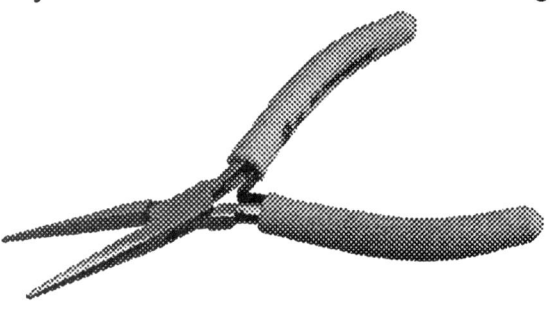

Figure 2-10 Long Chain Nose Pliers
(Reprinted by Permission of GC/Waldom Inc.)

Electricians Pliers

Electricians pliers are also called lineman pliers (**Figure 2-11**). They are used in heavy electrical work such as in residential or commercial construction. They have a cutting blade and flat jaws to bend heavy wire. They come in large and small sizes with insulated or uninsulated handles. The size dictates where they are used. They are also rugged in construction which can lead to abuse. Improper use of the pliers will damage them.

Figure 2-11 Electricians Pliers
(Reprinted by Permission of GC/Waldom Inc.)

Diagonal Cutting Pliers

Diagonal cutting pliers are used to cut wire or component leads (**Figure 2-12**). There are many styles of diagonal cutting pliers, with the jaws having a variety of designs and shapes allowing for different cuts. When choosing a pair of diagonal cutters, remember to choose a good quality pair. If they are to be used for multiple tasks, choose a pair with jaws that will cut a variety of wires without breaking. It is sometimes advisable to have many different sizes and styles available. In electronic repair, prepare for the unexpected.

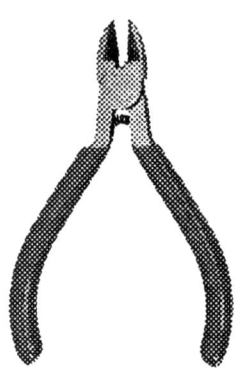

Figure 2-12
Diagonal Cutting Pliers
(Reprinted by Permission of GC/Waldom Inc.)

Nail Nippers

Nail nippers are a convenient and inexpensive tool to be used in an emergency. They can be used to remove insulation or trim small diameter wire or leads (**Figure 2-13**). They are not a replacement tool for diagonal cutters or wire strippers. There may be times when the proper tool is not available; therefore, it is good to have knowledge of other devices that can be used as substitutes when an emergency arises.

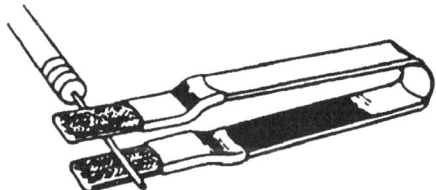

Figure 2-13 Nail Nippers

2-5   Component and Circuit Board Cleaning Tools

Component Lead Cleaner

This is a spring-actuated tool with lightly abrasive material inside the jaws (**Figure 2-14**). It is used to remove oxidation from component leads and/or wires. There are times when oxidation is so great that the flux in solder will not remove it. Oxidation occurs when the copper reacts to the oxygen in the air and forms an oxide coating on metals. Oxides prevent solid electrical connections when connecting wires or soldering components to printed circuit boards. It is sometimes essential that the oxide coating be physically removed prior to connecting the electrical parts.

Figure 2-14 Component Lead Cleaner

Typewriter Eraser

A typewriter eraser is a simple tool used in cleaning oxidation from traces on some printed circuit boards or terminals prior to soldering (**Figure 2-15**). Traces are the thin copper strips attached to the printed circuit board to connect circuits. However, caution should be exercised when using this tool on printed circuit boards as it can remove some traces due to its abrasiveness.

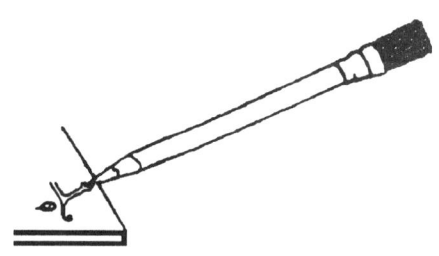

De-Soldering Tools

Solder Sucker

A solder sucker is a rubber bulb with a Teflon tip inserted in the end. It uses a vacuum to remove melted solder (**Figure 2-16**). It stores the cooled solder in the bulb section. This type of de-soldering tool is difficult to master, as it requires the operator to perform two tasks at once – melting the solder with the iron and squeezing the bulb in the melted solder and releasing it. The advantage to this tool is that it has no moving mechanical parts and it can be used over and over without wearing out. The disadvantage is that it requires considerable practice to master.

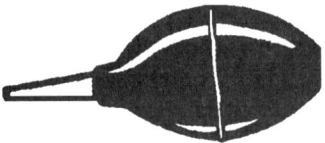

Figure 2-16 Solder Sucker

Solder Pullit

The solder pullit is a tool that uses the same principle as the solder sucker (**Figure 2-17**). It provides a vacuum to remove the solder. The pullit is loaded by depressing a spring connected to a piston, with a convenient release button. When the button is depressed, it releases the piston which creates a vacuum and pulls the melted solder into the chamber. The disadvantage is that plastic tools generate a static charge which can damage some small components. It is recommended that a metal pullit be used over the plastic type. This type will not generate a static charge. The pullits have moving parts which will wear out and require replacing. The piston uses a rubber "O" ring that requires regular cleaning and lubrication. The tips are made of Teflon and will not burn. This is normally the tool of choice. Proper care will extend its life. This tool is easier to use than the solder sucker and requires less dexterity.

Figure 2-17 Solder Pullit

Solder Wick

Solder wick is made of a copper braid impregnated with rosin (**Figure 2-18**). This material, as the name implies, absorbs liquid solder through capillary action. The wick acts like a paper towel absorbing the liquid. Remember that once a paper towel is used, it cannot be reused. The same holds true for solder wick. Another disadvantage is that it must be as hot as the soldering iron to work. The operator must exercise caution to prevent burns. The advantage is that it can be used in tight places. As far as cost, it is very expensive to use. The technician should use it wisely. It takes a certain skill to use solder wick; therefore, the technician should practice on unrepairable boards prior to using solder wick on connections requiring quality workmanship.

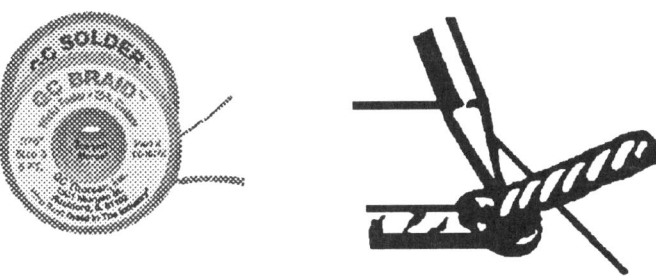

Figure 2-18 Solder Wick
(Reprinted by Permission of GC/Waldom Inc.)

Summary

When making electrical repairs, it is necessary to perform a quality job. It is important that the technician use the proper tools to perform repairs. There are a large variety of tools available and the proper tool to use is up to the technician. It is necessary to point out that there are a multitude of tools ranging from very cheap to very expensive. The choice is usually determined by the availability of funds. Tools purchased at the low end of the cost spectrum usually require replacement after being used a few times. This is normally due to the steel quality in the less expensive tools. It is recommended that the technician invest in a quality tool. Not only will it last a long time, but it will aid the technician in performing a quality repair.

As a technician, repairing damaged equipment or cables is part of the job. Proper tools will aid in making those repairs. If technicians are able to choose their own tools, it recommended that good quality tools be chosen. It is also recommended that tools be used properly and, after use, be cleaned and placed in a proper storage container. If tools are properly cared for, they will last a long time. Good tools will aid in performing a quality job.

2-6   Soldering Materials

Solder

Solder is used for fusing two metals together to form electrical connections. The most commonly used solder is a combination of **60% tin** and **40% lead** with a rosin core. The center core consists of a rosin flux that is used to clean the metal of oxides (oxides prevents the solder from fusing the two metals). (Note: The term flux refers to a metal cleaner and not an actual chemical.) The core of solder can contain one to five rosin cores (**Figure 2-19**). With the additional cores, the metal cleaning is achieved more quickly. Solder also comes in a variety of diameters. The technician determines the diameter and type of solder used. Solder and soldering will be covered in Chapter 3.

 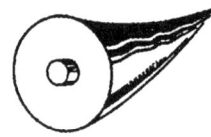 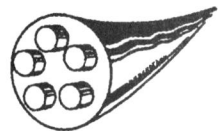

Figure 2-19 Rosin Cores

Rosin

Rosin is used for ensuring a good electrical connection by cleaning and wetting all surfaces during soldering. When making repairs, the rosin in the solder is normally sufficient to clean the surface area (rosin core solder) (**Figure 2-19**). If the rosin in the solder is not cleaning the surface during soldering, then use liquid flux or solder paste. Remember: excess flux must be removed after soldering, as it will attract dust particles, which can cause shorts in the electrical circuit.

Cleaning Chemicals

Isopropyl alcohol is a safe chemical that is used in removing oil and grease from the metal surfaces before soldering. It can also be used to remove the excess rosin after soldering. The advantage of this chemical is that it can be used safely because is it is not harmful to many soft materials, and it removes the majority of the rosin. The disadvantage is that isopropyl alcohol is not a total cleaner and can leave a sticky residue. This can be overcome if a bath tank is used for cleaning. This is not practical in most work areas, as it requires a special ventilation system to remove the fumes. It is safe for small areas. When soldering in a **WELL-VENTILATED** area where bench work is the norm, a flux remover containing methylene chloride, isopropyl alcohol, and trichlorotrifluorethane can be used. It should be noted that methylene chloride appears in section 313 of the toxic chemical list of Title III of the (SARA) Act of 1986 in the state of California as a NTF Anticipated Human Carcinogen. It is not on the FDA list for cleaning. Note: care should be taken when using this type of product. For safety's sake, consult the Materials Safety Data Sheet (MSDS) before using any type of flux cleaner.

Re-Insulating

After the connection is repaired and if it is a wire splice and not a component on a PC board, then it must be re-insulated. Normally wires are re-insulated with electrical tape. It should be noted that electrical tape is voltage rated. There are other materials that can be used to re-insulate.

One is spaghetti sleeving. This material is soft and should be used on low voltage wiring and where vibration is at a minimum. Spaghetti tubing is also made of hard plastic. This type is subject to sliding off the wire connection. If there is excessive vibration, electrical shorts can occur as the material is easily damaged.

A second is heat shrink insulation which is a material that, when heated, will form around the connection, preventing shorts. It is able to withstand vibration and abuse and still insulate properly. The disadvantage is that the tubing must be close to the size of the wire to be insulated. This requires the technician to carry a variety of sizes to use in making repairs.

A third is heat shrink tape which is another material that is more versatile in making repairs to broken cables. It is wrapped around the wire, then heated enough to shrink and form around the connection.

Summary

There are many materials available to the technician to aid in the proper repair of cables and equipment. The choice is up to the manufacturer of the equipment. If there are no such requirements, it is up to the technician to choose one that will provide the most professional repair.

SECTION I – WIRING AND SOLDERING

CHAPTER 3 – SOLDERING TECHNIQUES

Introduction Solder is a metal alloy used to join one or more metal surfaces together. An alloy is a mixture of two or more base metals. Most alloys contain large amounts of one metal, called the base metal, and smaller amounts of other metals. This joining process is called fusing. Fusing occurs when the mixtures are joined together by heat. When completed, a new alloy is formed. Soldering combines the metals of copper, tin, and lead. With the proper amount of tin and lead, the new alloy becomes very strong and versatile. It can also be used to mend metal objects. To be effective, the solder must melt more easily than the metals to which it is applied.

Soldering is the oldest known technique for joining metals that dates back to ancient Egypt.

Soldering is critical in today's repairs. Many new pieces of equipment require a skilled technician to locate problems and then, when possible, repair them through soldering. The technician should be as skilled in soldering as in locating the fault.

The skill of soldering requires an understanding of the makeup and characteristics of solder as well as the techniques of soldering.

Objectives After completing this chapter, you should be able to:

♦ Identify the proper tools used to remove solder.

♦ Determine the proper size and type of solder needed to make repairs.

♦ Determine the proper materials and cleaner used to make repairs.

♦ Understand the purpose of soldering and the characteristics of solder.

♦ Determine a properly soldered connection.

♦ Identify the causes and defects of an improperly soldered connection.

♦ Determine the proper procedure needed to prepare the metal surface before soldering.

♦ When required to solder, determine the proper soldering iron, solder, and procedure needed to properly solder a wire to a terminal.

♦ When working with stranded wire, perform the proper procedures needed to prepare the stranded wire for use.

3-1   Soldering Technology

Purpose of Soldering

The purpose of soldering is to provide a <u>reliable</u> electrical path for electricity to flow. The key word is reliable. When metals are joined without the fusion process, changes in temperature can cause the metals to separate. Another reason for soldering is that it provides <u>good contact</u> between the two surfaces. Proper soldering will prevent electrical contacts from separating during temperature changes. Electrical circuits are subject to excessive vibration. Without properly soldered connections, the circuits could become intermittent, <u>lose conductivity,</u> and the exposed copper will oxidize and corrode.

Characteristics of Solder

Solder is not a pure element but an alloy. It can contain a combination of elements. Some of those elements can be silver (ag), antimony (sb), bismuth (bi), and copper (cu). (The letters in parenthesis are the chemical symbols for the elements.) The two dominant elements in solder are tin and lead. The chemical symbol for pure tin is (sn) and its melting point is 450°F. The chemical symbol for lead is (pb) and its melting point is 621°F. Solder types are referred to by their percentage, or ratio, of **TIN** to **LEAD**. A 50/50 solder would mean that the ratio is 50% tin and 50% lead. The most commonly used solder for repairs is 60/40 which means 60% tin and 40% lead. This solder is stronger and more versatile for most soldered repairs. **NOTE:** This does not mean that other solders cannot be used, but 60/40 is more versatile.

The unique feature of solder is the melting point. When the two elements of tin and lead are combined, the melting point changes. As was stated earlier, tin melts at 450° and lead melts at 621°. When combined in a 60/40 solder, the melting point changes to approximately 375°F. It is generally accepted that all soldering takes place between 361°F to 576°F. Another characteristic of solder is the fusion process. Fusion occurs when the solder molecules combine with the copper molecules to form a bond. In the fusion process, there is a time factor, that is, the time it takes the solder to change into various states. This depends on the ratio of tin to lead. During the fusion process, solder melts in various stages. All solder except 63/37 melts in three stages. First it is a solid, then becoming plastic (soft plastic), and finally turning to a liquid. Solder then solidifies in reverse order, depending on the tin/lead ratio. The 63/37 changes from a solid to a liquid without going through the plastic stage. It is referred to as Eutectic solder, which means the lowest possible melting point of an alloy. The lowest melting point of solder is **361°F**. Therefore, when choosing solder for repairs, check the temperature requirements of the solder before soldering. It is important that the soldering iron selected will produce enough heat to melt the solder onto the surface area.

Figure 3-1 is a fusion chart for solder. Across the top of the chart are the tin and lead ratios. To use the chart, locate the tin to lead ratio across the top, then follow the line down until it touches the "**plastic**" line. (This is the point where the solder changes to a liquid.) Next follow the horizontal line to the temperature for that solder. **NOTE:** the 63/37 solder line intersects at the lowest portion of the chart at the letter "C". The actual temperature for 63/37 is 361°. The line at "C" indicates the division of the temperatures from 350° to 400° and it is 361°.

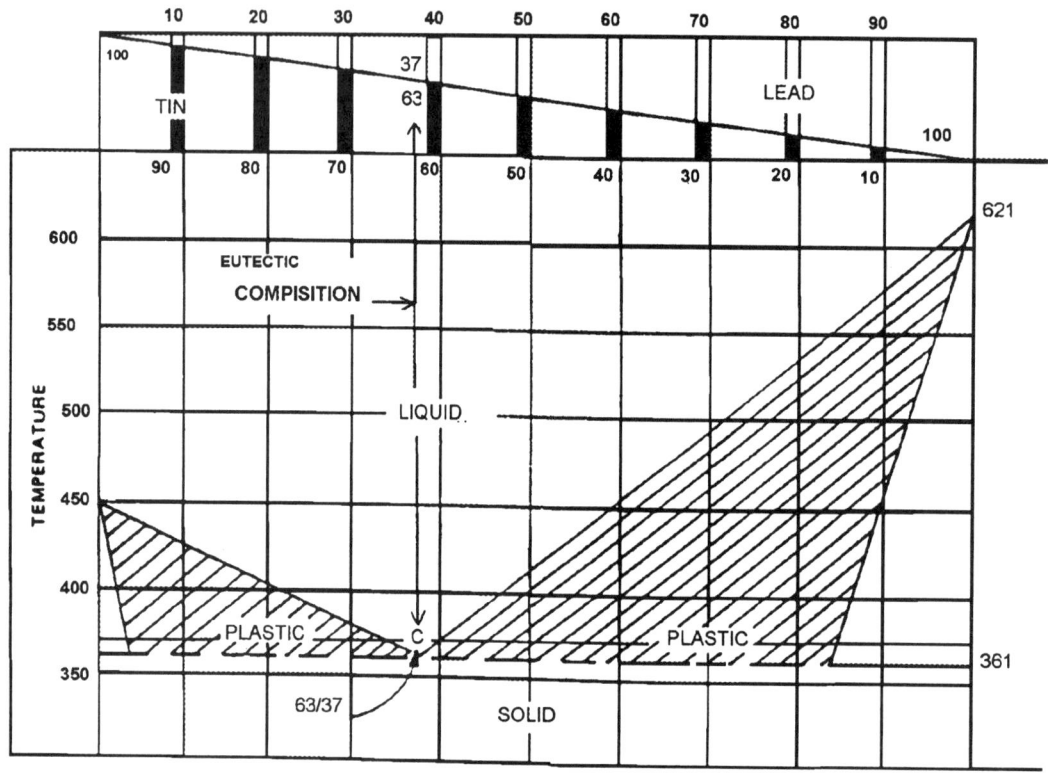

Figure 3-1 Fusion Chart for Solder

Why is it important to be familiar with this chart or melting temperatures of solder? When soldering, the temperature of the iron should match the melting point of the solder, as this will aid the fusion process. If the iron is too hot, it will cause the connections to become too hot and will burn the surface area, which can cause damage to the object being repaired. If it is not hot enough, the solder will not flow to the surface area and fuse the connection. The connection can have an intermittent condition. This can result in a cold solder connection.

Solder Types

There are many types of solders available for repairing circuits. The 60/40 is the most commonly used in electronic circuits. The 60/40 is also the most reliable and the strongest for electronic circuit repair. The 63/37 is soft for most electronic circuits that are moved around. Select the ratio that will best suit the job. The 63/37 solder is used in most solder baths. (Solder bath is a term that is used to describe the soldering of printed circuit boards.)

The ratio and the diameter of solders are not related. That is to say that most types of solders come in many different diameters. The recommended solder to be used for soldering the wires constructed earlier in the wire splicing lab is 60/40 with a diameter of .062. This solder will provide the proper amount of solder and time on the connection.

Not only is the solder type (RATIO) important in solder selection, but the diameter is also important. The diameter selection should be based on the type of connection to be soldered. If too large a diameter of solder were to be used to solder a computer chip in a printed circuit board, there would be too much mass flowing at one time. This could result in the bridging of two pins at one time or solder flowing across two of the printed circuit tracks. This can result in a short circuit. (A short is a connection of the wrong circuits together.) If a diameter of .031 were to be used, then the flow to the connection would be reduced to half. This would result in the computer pin being soldered properly. When selecting a solder diameter, it is up to the technician to select the proper diameter for the job. The two sizes mentioned, .062 and .031, should cover the majority of electronic repairs.

Note: Some circuits require certain solder ratios and diameters. If so, compliance is necessary. Some wires and other devices such as turret terminals may require larger amounts of solder. In this case, using the .062 diameter would be a good choice. When soldering, choose the proper diameter of solder and the proper wattage iron. This will aid in making a quality connection.

There are certain circuits that require a harder solder. Silver solders are harder. Silver is added to solder as a hardener and also is used to reduce arcing between connections. When silver is added to the tin-lead, it will raise or lower the melting point, depending on the quantity of each. Before attempting silver soldering, consult the manufacturer's data.

Along with the fusion characteristics of solder, the mass and characteristics of the connection should also be taken into consideration before soldering.

If a smaller diameter solder is used to solder a large connection, then it will require a longer time on that connection. The longer you hold the hot iron on the connection, the greater the risk of damage to the device being soldered. When a larger diameter solder is used, then too much solder can result. It is important to select the proper diameter of solder and the proper soldering iron before soldering. Proper iron selection will aid in reducing damage to the connection when soldering.

To summarize, before soldering, it is important to select the proper soldering iron that will produce the correct amount of heat, and the proper ratio and diameter of solder. This will reduce the time on the connection, thus preventing damage to the connection during the soldering process.

Soldering is both a science and an art. The science comes from knowledge of solder characteristics. The art comes from practice, practice, practice. The more you practice, the better you will become, and it will result in a highly reliable electronic circuit.

3-2   Critiquing Soldered Connections

When soldering connections, there are certain characteristics that a quality solder joint should have. There are two ways the technician can make that determination. One way is to X-ray the connection. As a repair technician, it is difficult to carry an X-ray machine in a toolbox. The second way is by visual inspection. When soldering, a visual inspection by the technician is the most practical. The technician should be able to determine the quality of the connection by a visual inspection using a good light.

What To Look For When Soldering

A good quality connection should be bright and shiny in appearance, like a new nickel, and it should properly cover the connection.

If a connection is incorrectly soldered, the defects will be visible. There are certain causes and effects of the poorly soldered connection.

Defects and Causes

The defects can easily be detected and, after detection, be repaired. When performing a visual inspection, if the connection is porous (pockmarks), having pits in the metal surface, it is considered defective. Impurities or dirt on the metal surface usually causes this. The metal surface area should be properly cleaned and the connection re-soldered.

It should have a good filling between the connections. After soldering, there should be no bare copper showing on the splice. The surface temperature not being hot enough usually causes this. There may have been insufficient heat from the soldering iron, or maybe the wrong wattage iron was used. This prevented the surface temperature from rising to the melting point of solder. To repair, re-solder the connection. This time be sure to get the surface temperature hotter. Check the wattage of the soldering iron and apply enough solder to cover the connection. Completely covering the connection with solder, aids in preventing oxidation and corrosion of the connection.

NOTE: *DO NOT OVER SOLDER THE CONNECTION. REMEMBER, THE CONNECTION MUST BE VISIBLE. TOO MUCH SOLDER PREVENTS PROPER INSPECTION OF THE CONNECTION.*

A quality job should have good adhesion and show no breaks in the soldered connection. Breaks in the solder connection can result in an intermittent condition. The connection should not have an excessive amount of solder or flux. Excessive amounts of solder can prevent the determination of the quality of the solder joint. Excessive amount of flux (which is sticky) can result in the connection collecting dirt, which can result in shorts. (A short is an accidental connection between two circuits, which can damage the electronic circuitry.)

The contour of the soldered connection must be visible. This pertains to printed circuit boards, when the leads extend through the board. The soldered connection should resemble a volcano, with the lead extending through the solder (**Figure 3-2**). The connection should NOT look like a Hershey's® candy kiss. That type of solder connection prevents inspection of the connection, and it also hides cracks under the base of the connection (see **Figure 3-3**).

   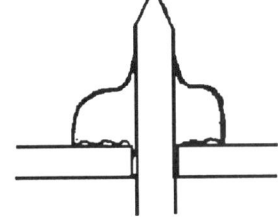

Figure 3-2 Good Connection   **Figure 3-3 Poor Connection**

When soldering, too much heat applied to the surface area causes the connection to look dull in color and/or rough, grainy, crusty, or wrinkled in appearance. That means it can exhibit any or all the stated conditions. This can be repaired by cleaning the surface and re-soldering the connection, spending less time applying heat to the connection.

When soldering a connection, if the surface temperature is not high enough to melt the solder properly, it can result in a cold solder joint. A cold solder joint can exhibit the same type of characteristics displayed by excessive heat. The major distinction is that it becomes more likely to appear dull and grainy looking. The end result could be an intermittent electrical circuit. What happens with the cold solder joint is that solder has not been allowed to pass through all melting stages and then allowed to solidify in the reverse order. **Figure 3-4** provides some examples of poor solder connections.

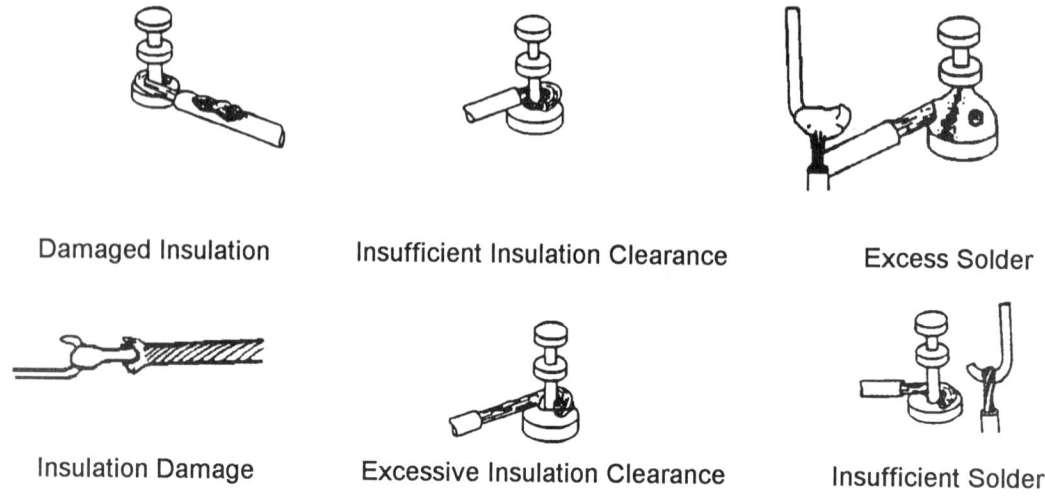

| Damaged Insulation | Insufficient Insulation Clearance | Excess Solder |

| Insulation Damage | Excessive Insulation Clearance | Insufficient Solder |

Figure 3-4 Poor Solder Connections

Figure 3-5 demonstrates the proper way to solder a turret terminal. The techniques are the same when soldering any type of connection.

They are as follows:

- Place the iron on the connection.

- Apply a small amount of solder to the iron to start the wetting action. (The wetting action occurs when the rosin in the solder cleans the surface area permitting the solder to flow freely to the tip.)

- Remove the solder from the iron and apply to the opposite side of the connection. The heat will cause the solder to flow around the connection to the soldering iron.

- When a sufficient amount of solder has melted, remove both the solder and the soldering iron.

- Allow the connection to cool without disturbing the connection. Inspect the connection under a light. If it is required, re-solder the connection.

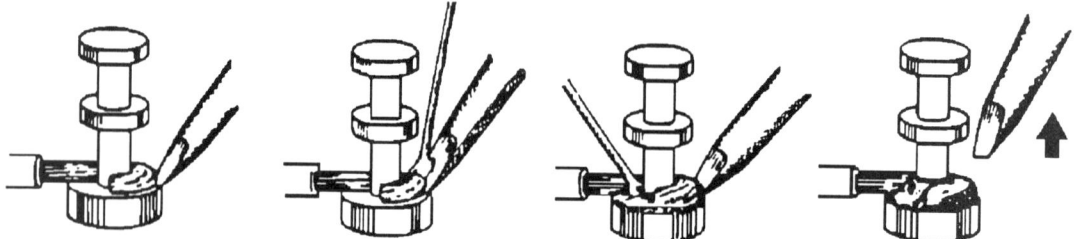

Figure 3-5 Good Solder Procedure

Summary

When soldering, the most practical way to determine a good solder connection is by visual inspection. This comes through practice and experience. Soldering is an essential skill in electronic repairs. Technicians should practice using scrap equipment until they become proficient and confident enough that, when required, a quality solder connection can be achieved without damaging the surrounding circuits. It is better to practice on something old than on something new. A good way to practice soldering skills is by soldering the wire splices mentioned in an earlier lab. The wire splices will provide the technician with some practice materials which, if mastered, will aid in soldering quality connections. When the technician is confronted with a connection requiring quality work, the connection can be soldered correctly the first time.

3-3   Metal Cleaners

Flux is defined as a cleaner for all metals. Different metals use different cleaners, for example, Borax is used as a cleaner in braising. However, in soldering, there are many types of flux cleaners.

Flux Classifications

Chloride, which has an acid base, is made of inorganic salts and is considered the most active of the flux group. It is effective on all common metals except aluminum and magnesium. It is not suitable for electronic circuits because it is highly corrosive and electrically conductive. It is also difficult to remove.

Organic flux is less corrosive but is as active as inorganic fluxes. This type is easier to remove than inorganic. It is also unsatisfactory for electronic circuits, because it must be completely removed to prevent corrosion. It has an extremely short lifetime at high temperatures. At high temperatures, the flux breaks down and does not clean the surface.

The previously mentioned flux cleaners should not be used on any electronic circuit as they are too difficult to remove. If they are not removed properly, the electronic circuits will corrode.

Rosin/Resin

This type of cleaner is ideal for soldering electronic circuits due to its molecular structure. The most common flux used in electronic soldering is a solution of pure rosin dissolved in a suitable solvent.

This solution works well with solder dipped metals and, although they are inert at normal temperatures, they break down and become <u>highly active</u> at soldering temperatures.

This makes rosin noncorrosive, except at soldering temperatures where cleaning the metal surface is essential. Rosin is also nonconductive making it ideal for soldering sensitive electronic circuits.

Resin fluxes are rosin based with additives that change some of the characteristics of pure rosin flux.

Most solder is used in electronic repairs are made in wire form with one or more cores of rosin flux. Some have many cores which aid in cleaning the metal surface faster.

When the joint or connection is heated and the wire solder is applied to the joint (not the iron), the flux flows on the surface of the joint and removes the oxide. This aids the wetting action of the solder (Figure 3-6). With enough heat, the solder flows and replaces the flux.

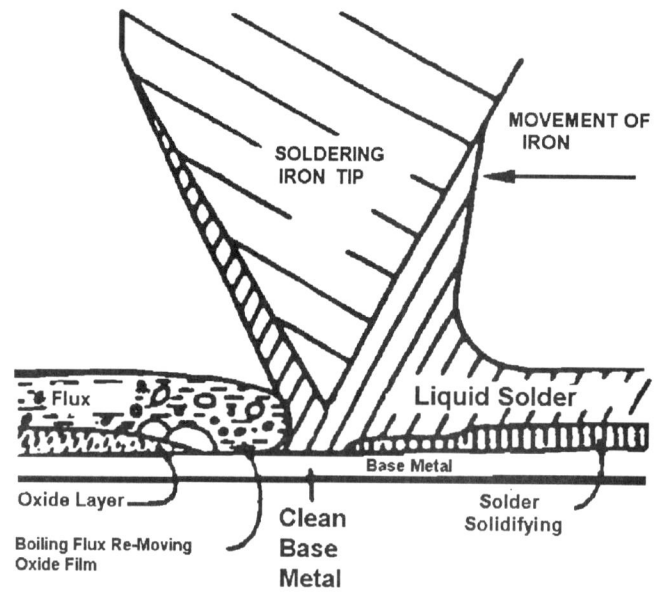

Figure 3-6 Solder Action

Insufficient heat during the soldering process results in a poor connection because the solder does not replace the flux. In addition to removing oxides, a good soldering flux must also help form the inner metallic solder bond by improving the solder's wetting action. The wetting action lowers the surface tension between the metals being soldered and the liquid solder. The wetting action also aids in carrying away the loosened oxides from the solder path. Flux is necessary in attaining the highest reliability in electronic solder joints.

Flux Cleaners

Normal flux cleaners have a combination of chemicals such as methylene chloride, with trichlorotrifluoroethane and isopropyl alcohol. These types of cleaners are effective on electronic circuits. Spray cleaners do not clean properly. Due to their fast evaporation rate, they tend to leave a residue.

Isopropyl alcohol is a good cleaner, and is a safer cleaner to use in closed areas. However, it takes a great deal more alcohol to clean the same area and it too leaves a residue.

The most effective cleaner would be methylene chloride, with trichlorotrifluoroethane and isopropyl alcohol in liquid form. The soldered area is scrubbed with a small nylon brush. Remember that the circuit is metallic and scrubbing will not damage the circuit.

NOTE: CARE SHOULD BE USED WHEN WORKING WITH CHEMICALS. CONSULT MATERIAL SAFETY DATA SHEETS SUCH AS MSDS AND OSHA CODES WHEN WORKING IN ENCLOSED AREAS.

Summary

Soldering requires not only skill but also an understanding of metals, their characteristics, and their reaction to temperature. The technician must be able to recognize the causes of poor quality connections and be able to correct them. It is important to recognize the characteristics of different solders, and know where and when to use them. The technician should also be aware of flux cleaners and which one to use or not to use. Some cleaners are corrosive to electrical connections and they should not be used.

Soldering is both an art and a science. To become a good solderer it is important to practice soldering on equipment and wires before attempting it on quality products. A good understanding of solder and soldering techniques coupled with practice will result in reliable electrical connections.

3-4   Soldering Procedure Labs

Lab A – Wires to Connectors

The following lab is to be used when soldering wires to a terminal connector.

1. Select the proper soldering iron for the job, depending on:
   a) Size of the connection
   b) Heat sensitivity of the components
   c) Proximity of other connections and wires

2. Heat and tin the soldering iron.

3. Strip and tin wires to be soldered.

4. Clean all surfaces to be soldered.

5. If the electrical connection is to be insulated, slide approximately one-inch length of spaghetti, or shrink tubing onto the tinned wire.

6. If possible, mechanically connect tinned wire to the terminal or lug by means of long nose pliers making sure the distance between terminal and wire insulation is no more than 1/32 to 1/8 inch, depending on the wire gauge.

7. If a mechanical connection is not possible, make sure the component to be soldered is held stationary in a holding fixture or clamp to prevent movement during soldering.

    NOTE: When soldering, the surface temperature must be greater than the melting point of solder, otherwise the solder will not fuse into the metal. Also, soldering is the fusion of metals, not gluing two surfaces together.

8. Attach the heat sink as close as possible to the connection without interfering with soldering operation.

    NOTE: The heat sink is used in both desoldering and soldering. In desoldering, it is used to protect suspected bad parts until verification of the failed part can be determined. In soldering, it is used to prevent overheating components.

9. If rosin-core solder is not used, brush a small amount of rosin flux on terminal.

10. While making sure that no part of the connection moves, apply a small amount of solder to the tip of the iron. This will create a wetting action, which allows the solder to flow freely over the connection. Then apply the hot iron tip to terminal or connection. As soon as solder has flowed freely over, around, and through the connection, remove solder and iron.

11. After the solder has cooled, clean connection with isopropyl alcohol or flux cleaner.

12. If connection is to be insulated, slide insulating material over connection.

13. If applicable, remove component from fixture.

Lab B – Soldering Splices from Chapter 1

After preparation of the soldering iron, select the appropriate size solder for the size 24-gauge splices from Chapter 1.

The 60/40 .062 rosin core solder would be appropriate to solder the splices.

Select or prepare a chisel tip. This will provide a transfer of heat to the splice so the solder can flow around and fuse the splice together providing a good electrical connection.

Select a holding fixture that can hold the splice rigid. If the splice is disturbed during cool down, it will cause a poor connection, possibly causing an intermitting connection.

Select a cleaning device for the tip. A wet terry cloth rag rolled will allow the operator to clean any excess solder from the tip prior to soldering.

Practice soldering the splices in the order constructed. Solder all of each type before proceeding to the next splice.

Proceed to practice soldering.

The following procedure will prove effective.

1.   Clean tip of excess solder.

2.   Place a small amount of fresh solder on the tip. This will enable the solder to flow toward the tip.

3.   Firmly place the tip under the splice, starting at the insulated end while applying solder to the top of the splice.

4.   Move the tip and solder together towards the other end as fast as the solder melts observing the solder flowing around the splice. This will take from 1.5 to 3 seconds to happen. Taking longer will possibly melt the insulation.

5.   Check the splice if bare copper is showing and repeat the process. If too much solder is on the splice, clean the tip and place it under the area of excess solder allowing it to flow to the tip.

Note: Move the tip from the point to the back of the tip to remove the excess solder. Clean excess solder and repeat until all excess solder is removed.

This procedure should be practiced until the splice can be soldered properly the first time.

Be patient and practice observing the process.

Practice, practice, practice is the only way to become proficient at soldering splices. Scrap wire splicing and soldering provides an economical means to become proficient. The timing on and off the splices can be applied to most cable repairs, as well as soldering components on printed circuit boards. However, contact the manufacturer before attempting repairs on PC boards. Repairs could damage the board.

Soldering is both an art and science pay attention to the science behind soldering, then solder for art, make the connection look professional.

Lab C – Stripping and Tinning

This lab is for practicing removal of insulation and then treatment of stranded wires. The purpose of tinning stranded wire is to prevent the strands from fraying during the connection process.

1. Make clean cut at wire end by cutting off small length with diagonal cutting pliers.

2. Lay wire in proper slot of properly adjusted wire strippers and strip approximately 1" of insulation from wire end.

3. Care must be taken not to remove wire strands when stripping or to nick if using solid wire.

    Caution: If too small a slot in wire stripper is used, the stripper will cut off some of the strands or nick solid wires. This could result in an electrical connection of higher resistance and weaken the wire.

4. If wire strands have been separated, gently re-twist wire in same direction as original twist.

5. After stripping, if wires are dirty or oily, clean wire strands or strand with isopropyl alcohol and paper towel.

6. Place heat sink on wire immediately adjacent to wire insulation before tinning stranded wires or solid wire.

    NOTE: Tinning is the process of applying a thin coat of solder to a surface area.

7. Tin soldering iron tip as follows:

    a. Plug in unit and allow it to heat up.
    b. Wipe tip with paper towel, wet sponge, or terry cloth towel.
    c. As soon as tip will melt solder, coat tip with solder, wipe off excess.
    d. If rosin-core solder will not be used, wipe small amount of liquid or paste rosin flux on bare strands of stripped wire.
    e. Set hot soldering iron in holder, stand, or vise.

8. Melt small bead of solder on tip and slowly draw bare wire through solder bead from heat sink toward wire end; apply additional solder as needed.

9. Remove heat sink from tinned wire.

10. After solder has cooled, clean the tinned wire with isopropyl alcohol or flux cleaner and paper towel or a clean rag.

11. If flux has crystallized, removal can be done with a blunt instrument like a screwdriver.

12. Solid wire is tinned the same way, even though it will not fray. Tinning solid wire has several advantages. It prevents oxidation and it allows the soldering process to fuse at a faster rate.

3-5 Soldering Terms

Bare copper showing – insufficient solder, flux did not clean connection.

Breaks in the solder connection – not enough wetting action to cause fusing action.

Burnt insulation – excessive heat, too long on connection.

Cold solder connection – insufficient heat, not permitting the solder to pass through all three states of liquefying and solidifying.

Crimping – applying mechanical pressure to compress a sleeve-type or cup-type electrical terminal to ensure a good electrical connection between the sleeve and the conducting wires it contains.

De-wetted solder joint – insufficient cleaning or insufficient use of flux.

Disturbed joint – connection moved before solder solidified.

Eutectic solder – solder that changes from a solid to a liquid without going through the plastic stage.

Excessive solder – poor joint: too much solder applied.

Flux (Rosin) – a liquid or solid which when heated cleans and protects surfaces to be soldered.

Insufficient solder – too little solder applied.

Intermittent connection – too much solder on connection and cannot be properly inspected, too much rosin, collects dirt, causing intermittent shorts.

Land – printed wiring attached to the surface of a printed circuit board.

Overheating – causes the connection to appear dull in color, rough, grainy, crusty, and wrinkled in appearance.

Oxides – films and impurities which form on the surface of metals when exposed to air or water and which, if not cleaned off, will prevent a good bond between the surfaces and solder.

Porosity – caused by impurities in the metals, examples are dirt or oxidation.

Rosin – a material obtained from pine trees which is used during soldering to help ensure a good bond between the solder and the metal surfaces.

Rosin solder joint – excessive rosin or insufficient heat.

Solder short – excessive solder between connections--caused by carelessness.

Solder temperatures – soldering takes place between 361°F - 576°F.

Soldering – metallurgical method of joining metals through the use of a filler material called solder.

Stripping – removing insulation from electrical conductors.

Tinning – the application of a small amount of solder to the surfaces to be soldered, to help ensure good wetting during soldering, and to stranded wires to prevent fraying.

Wetting – the ability of molten solder to flow over and fuse completely with the metal surfaces to which it is applied.

SECTION II – HAND TOOLS AND HARDWARE

CHAPTER 4 – HAND TOOLS

Introduction   Basic hand tools are those tools that are basic to all types of repairs. They are also referred to as common tools. Tools are needed to make repairs. Proper tools will aid in performing the repair without damaging the equipment. Using tools improperly can damage the equipment beyond repair. Tool names can also denote their purpose and identify on which device it is to be used. Hand tools, if taken care of, will last a long time. High quality tools will perform properly and last longer than poor quality tools. Poor quality tools will break or deform from use. If a deformed tool is used, it can cause damage to the object under repair. It is important for the service technician to become familiar with the proper names of tools and where they are used.

Objectives   After completing this chapter, you should be able to:

♦   Identify tools by their correct name.

♦   Determine the proper tool for the job.

♦   Identify tools by their picture.

♦   Identify precision measuring tools.

4-1   Basic Hand Tools

Tightening tools are classified according to their ability to generate torque on a nut, bolt, or screw.

Torque -** **Force exerted by a twisting action, it is usually designated in inch/lbs. or foot/lbs.

Nut Driver – This tool is used to tighten nuts on threaded bolts or tightens nuts on studs (**Figure 4-1**). The nut driver is a low torque tool and has a small mechanical advantage. The tightening process is limited to the wrist strength of the operator. Nut drivers are used on nuts or bolts that range in size from 1/16" to 1/2". Good quality tools have color-coded handles for ease of identifying tool sizes.

Figure 4-1 Nut Driver

Open End Wrench – This tool is also used to turn nuts or bolts (**Figure 4-2**). This is a medium torque tool with a good mechanical advantage. The disadvantage is that it can slip or come off the fastener inadvertently. The reason for this is that it has only two surface points of contact. When excessive torque is applied, the four surfaces of the wrench and the bolt will not hold on tight. This allows the wrench to slip and can either damage the bolt or hurt the operator. Care should be taken when using this tool on extremely tight bolts or nuts.

Figure 4-2 Open End Wrench

Box Wrench – This is another tool used to turn nuts and bolts. It can be used as a high torque tool to loosen or tighten bolts or nuts. It does this by surrounding the fastener on all sides, which reduces the chance of it slipping off (**Figure 4-3**). The box wrench comes with a 6-point or 12-point end. The 6-point end is the better of the two for extremely tight fasteners. All hex fasteners, being 6 sided, allow for a tight fit between the two surfaces. A 12-point end on the other hand is made by dividing the 6 points in half thus 12 points. The advantage of this tool is that it can be used in areas where the space is close. It requires less swing of the wrench. If the bolt is extremely tight, it is recommended that the 6-point wrench be used to break it loose, then the 12-point can be used to finish removing it.

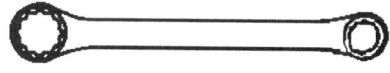

Figure 4-3 Box Wrench

Combination Wrench – This is a wrench with a box wrench on one end and an open-end wrench on the other. Usually both ends are the same size (**Figure 4-4**). The advantage of this tool is that the box end can loosen the fastener and the open end can remove it. The disadvantage of this wrench is that it usually comes in one size on both ends. This feature requires that more sizes of wrenches be part of a tool kit. In contrast, box wrenches have a different size on each end of the wrench. This feature reduces the number of tools that are needed.

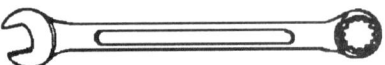

Figure 4-4 Combination Wrench

Ratchet Drive Wrench – This tool has an internal mechanical mechanism that is used to turn a socket, which attaches separately, and is used to remove semi-tight bolts or nuts (**Figure 4-5**). With the ratchet drive wrench, one has the ability to work in a confined area. The ratchet mechanism allows for a back and forth movement in a tight area. The ratchet mechanism can be used for reversing the direction by the flick of a lever. This makes it more versatile in removing semi-tight fasteners. Caution should be exercised when using the ratchet wrench.

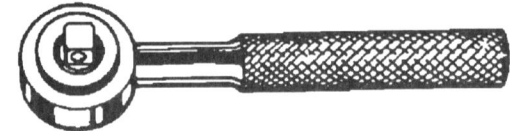

Figure 4-5 Ratchet Drive Wrench

The ratchet should never be used to break loose extremely tight fasteners. To do so may cause the mechanical mechanism to break, rendering the wrench useless. If used properly, the ratchet wrench will last a long time.

Socket – This tool looks like a short piece of pipe (**Figure 4-6**). One end has a square drive, with square sizes from 1/4", 3/8", 1/2", 3/4", and 1". The other end of the socket has either 6 or 12 points. The socket must be used in conjunction with a ratchet wrench or a breaker bar or other driving mechanism. The breaker bar, which looks like a "T", is stronger than the ratchet wrench. Therefore, the breaker bar should be used to loosen the fastener, then the ratchet drive is used to remove it. The socket can withstand extremely high torque, and it is used on the end of torque wrenches for tightening fasteners, which will be discussed later on.

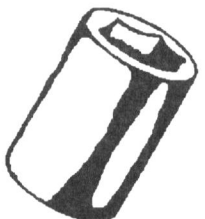

Figure 4-6 Socket

Allen Wrench – This is used to drive hex head or "socket" head bolts or set screws (**Figure 4-7**). It is considered a high torque tool. Allen wrenches range in size from 1/32" to 1/2". If a wrench is 1/32", it is still considered high torque. Tools that fit the hex head exactly are considered high torque no matter what their size. They can be purchased individually or in sets with straight shanks with handles or as sets that are mounted to a holder that allows for them to be folded down into a handle. They are used extensively on electro-mechanical equipment to secure knobs, pulleys, or gears to shafts with the Allen/socket head screws.

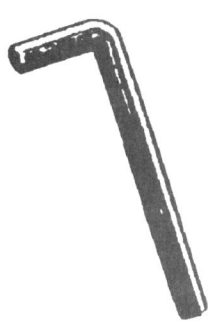

Figure 4-7 Allen Wrench

Tongue & Groove Pliers – This type of pliers is probably the one type that is called by more names than any others (**Figure 4-8**). It is used in many professions and each calls it something different. Some other common names are channel locks or water pump pliers. The correct name is tongue and groove pliers. They are used to tighten or loosen large diameter pieces of hardware that do not require high torque. These pliers have 4-6 size adjustments. The quality of this tool depends on the manufacturer. There are high quality tools and poor quality tools. It is advisable to invest in a good quality tool that will last longer.

Figure 4-8 Tongue & Groove Pliers

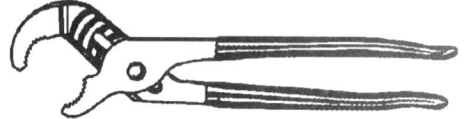

Slip Joint Pliers – This is a common tool found in almost everyone's toolbox. It is a very versatile tool that is used by all types of service-related industries (**Figure 4-9**). It has a slot to adjust for small or wide holding ability. The jaws are internally curved to grasp round objects. It also has a flat grooved end to aid in turning flat square objects. Care should be taken when purchasing this type of tool. If the steel quality is poor then the tool cannot be used for high torque items.

If a poor quality tool is used on high torque items, the groove area that is used in the adjustment will become damaged, and the pliers will not hold the item. This can cause damage to the item. With high quality steel tools, the slip joint will hold and not damage the device. *Quality tools do quality work.*

Figure 4-9 Slip-Joint Pliers

Long Chain Nose Pliers – This tool is one of the most misnamed of all the pliers (**Figure 4-10**). These pliers are *INCORRECTLY* referred to as needle nose pliers (Figure 2-9). The difference between the two is that the chain nose pliers have a large base that tapers evenly to the jaws. The base of the needle nose pliers is narrow and has a long thin taper to the jaws. Both are used to bend component leads or wires and many other applications. The long chain nose pliers are for heavier duty work and the needle nose pliers are for fine work. Incorrect usage can result in damage to the pliers.

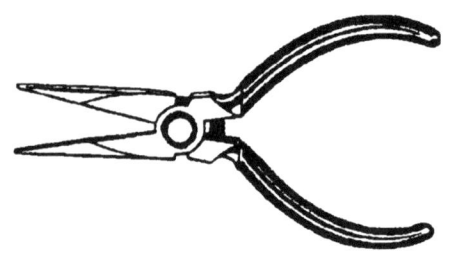

Figure 4-10 Long Chain Nose Pliers

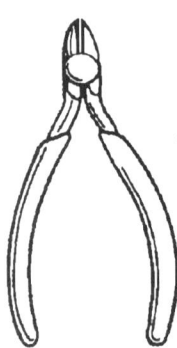

Diagonal Cutting Pliers – This is a versatile tool that is used in cutting wire or component leads (**Figure 4-11**). The tool has many different designs of the cutting jaws. The different angles and shapes allow the operator to make cuts on any angle. With the many different designed pieces of electromechanical equipment to be repaired, one tool will not work for all pieces. A good quality set of diagonal cutting pliers will last a long time with proper care and use. With good quality pliers, the cutting of the wires will not damage the cutting edge of the pliers. When using a poor quality set, the jaws are easily damaged due to the soft steel of poor quality tools.

What normally happens is that the materials being cut are harder than the pliers. Continued use of soft steel pliers results in the jaws loosing their cutting edge. When selecting pliers, choose a pair that will be the correct angle and are made from good quality steel.

Pin Punch – It is a thin tool used to extract or drive pins (**Figure 4-12**). The type of pins are roll pins, dowel, and tapered pins. The pins are normally flush with the surface area and are very tight in the hole. The pins are usually made of hardened steel that requires a hardened steel pin punch to remove them. The pin punches are made in various sizes, which must be matched closely to the pins. The punch is a long thin round metal with a flat tip. Incorrect size pin punches will normally damage the walls or the pins when trying to remove them. It is important to select the correct length pin punch so that when removing the pins an extension piece is not needed.

When using pin punches, be careful of the end of the punch being pounded. This end should not have a mushroomed head. If it does, remove the mushroom end with a grinder. The mushroomed head can break off and injure someone.

REMEMBER – SAFETY FIRST.

Inspection Mirror – This device is used to view areas in behind where it is impossible to see normally (**Figure 4-13**). It is better to use this device to look behind hard to see areas for damage than to remove many parts to only find that nothing is wrong. It is a very useful device to have and can be a real time saver.

Figure 4-13 Inspection Mirror
(Reprinted by Permission of GC/Waldom Inc.)

Pin Extractor – A tool used to remove the contact pins in electrical plugs (**Figure 4-14**).

Tru-arc Pliers – Also **called** snap ring pliers, they are for installing or removing tru-arc snap rings (**Figure 4-15**). The rings can be either internal or external.

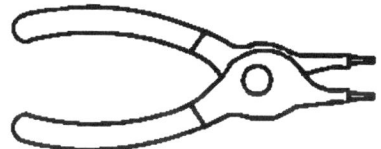

Wire Strippers – Wire strippers should be adjusted so that the insulation is cut and not the wire (**Figure 4-16**). A small nick will weaken or cause increased resistance in the wire.

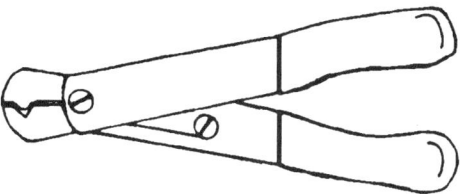

Dies – Dies are for cutting threads on round stock, e.g., threads on a bolt (**Figure 4-17**). There are two sides to a die, the starting side and the finishing side.

Taps – Taps are for cutting internal threads in most materials, so that a bolt or stud can be threaded into it (**Figure 4-18**). There are three types of taps, **starter, finishing**, and **bottoming**.

Abrasive Cloth – A cloth with an abrasive material glued to it (**Figure 4-19**). It comes in various grits: fine, medium, course, etc. This is also known crocus cloth or emery cloth.

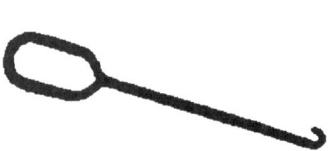

Spring Hook Tool – This tool is used to remove or install springs (**Figure 4-20**).

Feeler Gauge Set – A set of various thickness blades which is used to check various unknown blade thicknesses (**Figure 4-21**). Sets usually go from .015" to .025".

Torque Wrench – This tool is used to apply a specific force to a nut or bolt (**Figure 4-22**). The torque wrench ensures uniformity in tightening a series of bolts or nuts on a single application.

File Card – A file card is a short hair wire brush that is used to clean foreign material out of a file (**Figure 4-23**).

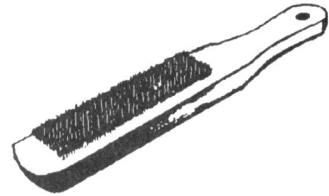

Screw Starter – A tool with a section in the center which twists against the inside of the flat or Phillips slot in order to hold it, to get it started, in a difficult location (**Figure 4-24**).

4-2   Fabricating Tools

Scribe – A long pointed metal tool used to mark on metal for drilling, cutting, bending, etc. (**Figure 4-25**).

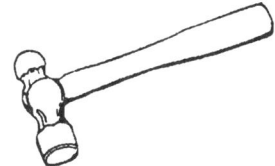

Ball Peen Hammer – A standard hammer used in metalworking. The ball peen hammer comes in various sizes and weights (**Figure 4-26**).

Center Punch – A hardened punch used to make an indentation in material prior to drilling (**Figure 4-27**).

Hacksaw – The hacksaw is used for cutting of metals. Blades come in fine, medium, and coarse (**Figure 4-28**). Hacksaws will be covered in **Chapter 8**.

Round File – It is sometimes called a rat tail file and is used for filing round holes (**Figure 4-29**).

Flat File – A common file used to smooth or remove material for proper fit (**Figure 4-30**).

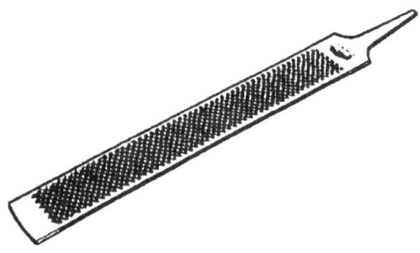

4-3 Precision Measuring Tools (See Chapter 6)

Dial Caliper – This tool is used to measure accuracy from approximately .001" to approximately 6" although calipers can be larger than 6". It is easy to use, read, and understand. It is also three tools in one; it reads inside, outside, and depth dimensions (**Figure 4-31**).

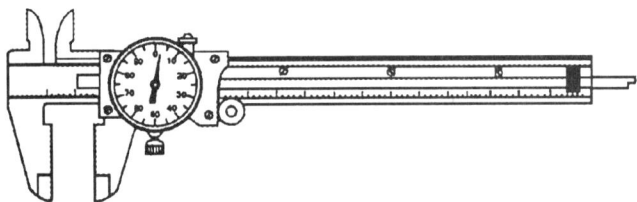

Dial Indicator – A tool used to measure out of round on a circular application or other application with variations of thickness (**Figure 4-32**).

Micrometer – The most accurate measuring tool for measuring outside dimensions (**Figure 4-33**).

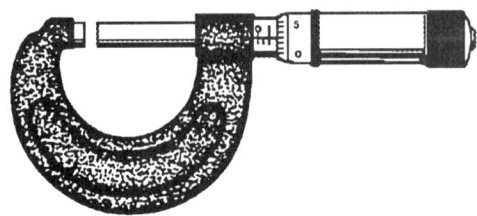

4-4   Hand Tool Safety

Hand tools are probably the most misused tools in industry. Basic tools have not changed for many years; however, there are various grades of hand tools available. The poor grade tools can be dangerous to the operator as they may break under stress and cause the operator to be injured. There are also medium quality tools and high quality tools. The technician should choose the one best for the jobs on which they will be used. Good quality tools should be well cared for, observing the hand tool safety rules. Properly used and cared for tools will last for many years.

Hand Tool Safety Rules

1.   Never use a screwdriver to check high voltage.

2.   Always select a screwdriver with the proper size head to fit the slot in the screw.

3.   Before using a hammer, check the head for a tight fit in the handle.

4.   Never use a file that is not fitted with a handle.

5.   Keep a hot soldering iron in its holder.

6.   Never clean soldering guns/irons by slinging solder.

7.   Replace all frayed or faulty power cords.

8.   Turn in all broken or worn out tools when they are discovered.

9.   Wear safety glasses/goggles when cutting wire, soldering, using the grinder, drill or solvents, or working with cathode ray tubes (CRT's).

10.   All tools that are not in use should be properly stored.

11.   Check chisels and punches for mushroomed heads before using.

12.   Do not cut or use tools with the working edge toward you.

13.   Allow hot soldering irons to cool before storing.

14.   Never pull on a line cord to disconnect a plug.

NOTE: The hand tool safety rules have been established to aid the user in preventing accidents. In most cases, when hand tools are used properly, a quality job results. A good set of hand tools, properly care for, is as important as the education used to diagnose the problems. As has been stated, with an education and a good set of tools, one can earn a good living.

Summary

Tools are the second most important things that the service technician must have. The first is the training to locate the problem, next to have the tools and skills to use them properly. The quality of the tools is important in that poor quality tools will result in poor quality repairs. A variety of tools is also important. It takes many different tools to make the variety of repairs that will be required while repairing the equipment. It is the responsibility of service technicians to take care of their tools and to have the proper tools to perform the variety of repairs that will need to be performed on electromechanical equipment.

SECTION II – HAND TOOLS AND HARDWARE

CHAPTER 5 – BASIC HARDWARE

Introduction   Hardware falls into a category that everyone takes for granted. It is also given little thought until something that was working, stops working, because a critical piece of hardware fell out and no one can find it. Everything made requires some form of hardware to hold it together. Automobiles can use as many as 3,500 pieces of hardware. Telephones can have as many as 73 fasteners. There are many companies that manufacture fastening devices commonly referred to as hardware.

This chapter is divided into four basic categories: Driving Recesses, Head Styles, Fasteners, and Fastening Devices.

Objectives   After completing this chapter, you should be able to:

- ◆   Identify the four basic categories of hardware.

- ◆   Recognize different types of driving recesses.

- ◆   Determine which type of head style works best for a given job.

- ◆   Identify four basic grades of fasteners.

- ◆   Identify fastening devices by name.

- ◆   Determine which fastening devices will work best in a given situation.

5-1 Driving Recesses

Slotted – For low torque driving, used where you're not worried about damage if the driver slips.

Phillips – More actual blade surface grips screw head; used where more driving torque is needed or when driving with a power drill.

Allen Head (Hex) or Socket Head – Most positive torque recessed drive, used mainly on set screws and counterbored screws for a flush fit and to reduce chance of surface damage.

Clutch – Security type of driving recess; prevents access by unauthorized personnel. Also used in high speed production, providing positive driving of hardware.

Bristol – Same uses as the Allen Head but requires a special tool to remove or tighten screw.

Hex Head – A six-sided head with corners trimmed to close tolerances, found on machine screws, cap screws, and machine bolts. Used where wrench tightening is desired or greater tightening force is needed.

5-2 Head Styles

Oval Head Screw – The bottom of the head countersinks and the top protrudes slightly, giving a decorative look. It is used where appearance is a factor.

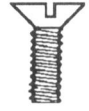

Flat Head Screw – This screw countersinks to flush with surface and is used where clearance is important and for appearance.

Truss Head Screw – A larger diameter head, used mainly in sheet metal to cover oversize holes. It is also used in soft materials for surface holding area.

 **Pan Head Screw** – This screw has a larger head surface and slot for greater driving power and to spread out load or cover larger hole. It is used mainly in metal fastening where a protruding head is specified.

 **Round Head Screw** – This screw protrudes above the surface and is used where work will be disassembled and appearance is not important.

 **Fillister Head Screw** – Usually with deep slotted drive, has a deeper head than the round head or pan and is used in counterbored holes where head driving power is needed.

5-3 Fasteners

 **G2** – Common steel bolt; every day use; cheapest to purchase and low strength.

 **G5** – Heat treated steel bolt; better grade, where more strength is required.

 **G8** – Case hardened, high grade steel; used where maximum strength is required; expensive.

 **Stainless Steel** – Used in applications where food is processed or corrosion is a problem.

 **Split Lock Washer** – Locking washer made out of spring steel; inexpensive.

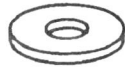 **Flat Washer** – Used to increase head diameter for better holding and to reduce surface damaged.

Internal Tooth Lock Washer – Used when good holding is required along with good appearance; found on the outside of chassis.

 External Tooth Lock Washer – Used when good holding is required but where appearance doesn't matter; usually found on the inside of equipment.

 Standard Nut – Commonly used nut, grade 2.

 Jam Nut – Similar to standard nut but is thinner; used to "jam" against standard nut as a locking method.

 **Tinnerman Nut** – It is used in areas difficult to reach such as covers or speaker backs. It is made of spring steel, usually flat, and has only one thread.

 **Castellated Nut** – Has a crown on top of it where a cotter pin or wire can pass through to lock it into position.

 **Wing Nut** – Nut used where it would be removed and replaced quite often. It can also be referred to as a butterfly.

 **Cap Nut** – Also called an Acorn nut; used where looks and safety are important.

 Stop Nut – Also called a Nylon nut. A self locking nut using a small ring of plastic or nylon in it to increase its resistance to loosen.

 Keeper Nut – Nut with an external tooth lock washer attached to it. Convenient to use in hard to reach areas.

 **Prong Tee Nut** – Used to fasten wood to metal.

 **Cotter Pin** – A common pin used to hold castellated nuts from moving.

 **Dowel Pin** – Solid steel pin used as an alignment or guide pin.

 **Roll Pin** – A spring steel pin generally used to connect a gear, pulley, or sprocket to a shaft.

 **Tapered Pin** – Solid steel pin, tapered and generally used to align two different surface areas.

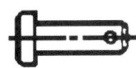 **Clevis Pin** – Used when two items are to be connected but need to move freely.

 **Square Key** – A long key used to drive a gear, sprocket, or pulley – 1/2 in the shaft, 1/2 in the gear, etc. Made of mild steel, will shear if the torque is exceeded.

 **Woodruff Key** – A driving device in the shape of a half moon – 1/2 of the key is recessed into the shaft and the other 1/2 extends into the gear, pulley, etc.

5-4   Fastening Devices

 Tru-Arc Outside – Used to hold something on the outside, such as a shaft.

 Tru-Arc Inside – Used to hold something on the inside, such as a bearing.

 "E" Ring – Used to hold a shaft in position.

 **"O" Ring** – Used to hold an object in a set position.

 **Thumb Screw** – Machine or screw threads with the driving end having wings for tightening or loosening by hand.

 **Lag Screws** – Large screws with hex heads used to hold wood together.

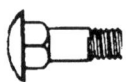 **Carriage Bolts** – Have a smooth head with a square section just under the head, used to connect wood to wood and wood to metal.

 **Self Tapping Machine Screws** – Machine screws, 10-24, 6-32, 8-32, etc. with slots on the beginning of the threads, that cut or form threads, so the body can screw into it.

 **Sheet Metal Screws** – Wood type screw threads used to connect thin sheet metal together.

Summary

There are many different pieces of hardware on the market. Many are stock items and others are specialty items. It is important to know the stock items by name. The manufacturer will identify specialty items. If a piece is a specialty item, it should be replaced with the same specialty item. There are many items that are interchangeable, but many are not. Putting the wrong replacement hardware in a device can cause damage. Some damage is not noticeable at first, but results in a malfunctioning piece of equipment which could cost more to repair then putting the correct hardware on in the first place. Do not reinvent the equipment...follow manufacturers instructions on replacement hardware.

SECTION II – HAND TOOLS AND HARDWARE

CHAPTER 6 – MEASURING DEVICES

Introduction   Precision measuring devices such as the micrometer, dial calipers, and dial indicator are useful tools to the technician. They are used to aid in modifications in making precision adjustments to mechanical devices. As mechanical devices become smaller and smaller, relying on tape measures for adjustment is usually out of the question. This chapter discusses three basic measuring devices with practice exercises to aid in familiarization with some of the tools. A thorough knowledge of the tools requires the technician to have the actual measuring devices on hand. The procedures for screw thread identification will work with practice. There are thread gauges available that also require practice to properly identify threads. The procedure for thread determination with a dial caliper works in conjunction with drill exercise chart in 7-4. Use the chart and procedure for determination of stock thread sizes. It should be noted that many of the precision measuring devices discussed also come in electronic design still requiring the technician a basic knowledge on their operation.

Objectives   After completing this chapter, you should be able to:

- Identify parts of a micrometer.

- Determine measurements when using a micrometer.

- Set linear measurements to a given dimension with a micrometer.

- Make inside, outside, and depth measurements with a dial caliper to within .001 of an inch.

- Determine the diameter of a bolt and convert the measurements to stock sizes.

- Set the dial caliper to a specific linear measurement and determine the number of threads per inch of a given bolt or screw.

- Identify the SAE (Society of Automotive Engineers) threads of a bolt or screw.

- Use a dial indicator to measure a shaft on a rotating device and determine if it rotates true (no out of round movement).

- Set the dial indicator to a specified linear measurement and measure the thickness variations of several materials.

6-1   Micrometers

The micrometer is a precision measuring instrument. It is used to measure outside dimensions to an accuracy of .0001 or 10 thousandths of an inch. Other micrometers are designed to make even smaller measurements. The micrometer, discussed in this chapter, will be used to measure .001 of an inch.

The 0.001-Inch Micrometer

A 0.001-inch outside micrometer (**Figure 6-1**) is shown with its principal parts labeled.

The part to be measured is placed between the anvil and the spindle. The barrel of a micrometer consists of a scale that is 1" long. The 1" length is divided into ten divisions each equal to 0.100-inch. The 0.100-inch divisions are further divided in four divisions each equal to 0.025 inch.

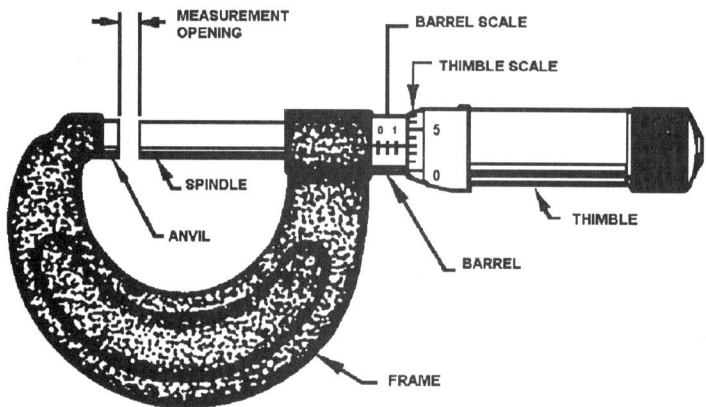

Figure-6-1 Micrometer

The thimble has a scale (**Figure 6-2**) which is divided into 25 parts. One revolution of the thimble moves 0.025 inch on the barrel scale. Therefore, a movement of one graduation on the thimble equals 1/25 of 0.025 inch or 0.001 inch along the barrel.

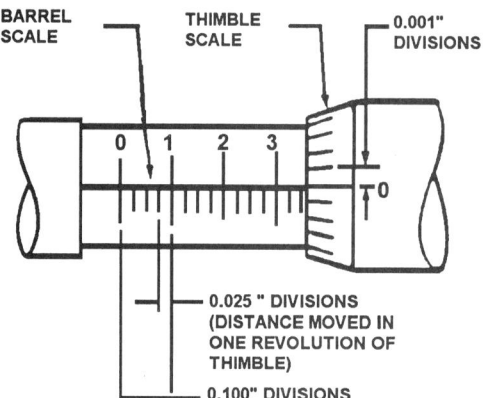

Figure-6-2 Barrel & Thimble Scale

Using the 0.001-Inch Micrometer

A micrometer is read by observing the position of the bevel edge of the thimble in reference to the scale on the barrel. Observe the greatest 0.100-inch division and number of 0.025-inch divisions on the barrel scale. To this barrel reading, add the number of the 0.001-inch divisions on the thimble that coincide with the horizontal line (reading line) on the barrel scale.

Procedure: To read a 0.001-inch micrometer

- Observe the greatest 0.100-inch division on the barrel scale.

- Observe the number of 0.025-inch divisions on the barrel scale.

- Add the thimble scale reading (0.001-inch division) that coincides with the horizontal line on the barrel scale.

Example: Read the micrometer setting shown in Figure 6-3 below.

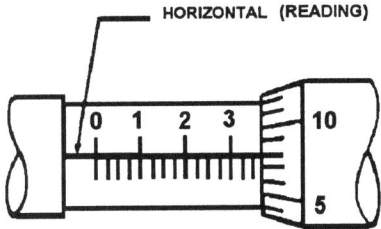

Observe the greatest 0.100-inch division on the barrel scale (3 × 0.100" = 0.300").

Observe the number of 0.025-inch divisions between the 0.300-inch mark and the thimble (2 x 0.025" = 0.050").

Add the thimble scale reading that coincides with the horizontal line on barrel scale (8 × 0.001"= 0.008").

Micrometer reading: 0.300" + 0.050" + 0.008" = 0.358" (ANSWER)

Linear measurement: Procedure to set a 0.001-inch micrometer to a given dimension:

Turn the thimble until the barrel scale indicates the required number of 0.100-inch divisions plus the necessary number of 0.025-inch divisions.

Turn the thimble until the thimble scale indicates the required additional 0.001-inch divisions.

Example: Set 0.949 inch on a micrometer (Figure 6-4)

Turn the thimble to nine 0.100-inch divisions plus one 0.025-inch division on the barrel scale (9 X 0.100" + 0.025" = 0.925").

Turn the thimble an additional twenty-four 0.001-inch thimble scale divisions (0.949" - 0.925" = 0.024").

The 0.949-inch setting is shown.

Figure 6-4

Exercise 1: Reading micrometer settings lab

Using a 0.001-inch micrometer

Read the settings on the following 0.001-inch micrometer scales (**Figures 1 to 12**).

1. .589
2. _____
3. .756
4. _____
5. .88
6. .413
7. .763
8. .623
9. .157
10. .715
11. .949
12. .520

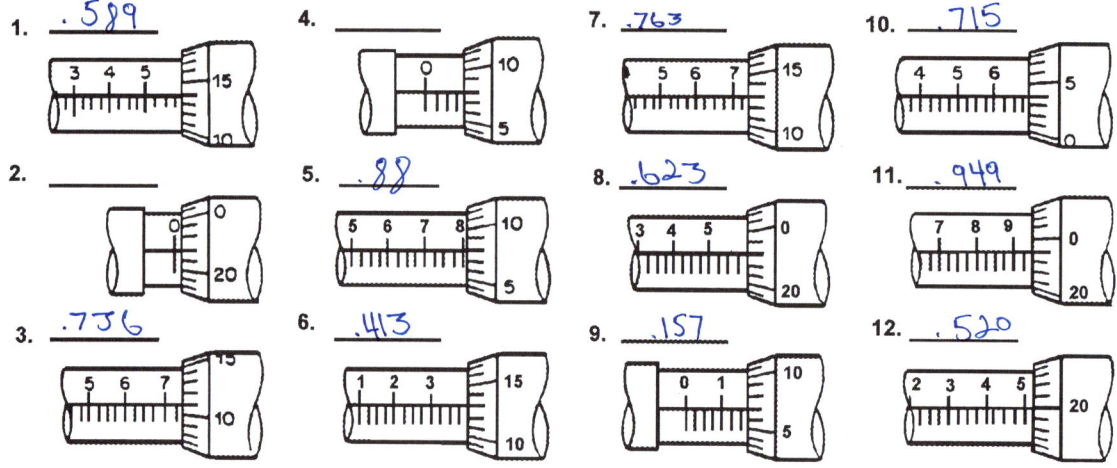

Exercise 2: Using pre-given barrel and thimble settings, determine micrometer reading.

Given the following barrel scale and thimble scale settings of a 0.001-inch micrometer, determine the readings in the tables. **The answer to the first problem is given.**

BARREL SCALE SETTING IS:	THIMBLE SCALE SETTING IS:	MICROMETER READING IS:
0.425" - 0.450"	0.016"	0.441"
0.075" - 0.100"	0.007"	.082
0.150" - 0.175"	0.003"	.153
0.875" - 0.900"	0.012"	.887
0.300" - 0.325"	0.024"	.324
0.000" - 0.025"	0.021"	.021
0.025" - 0.050"	0.013"	.038
0.750" - 0.775"	0.017"	.767

Summary

The micrometer is considered the most accurate precision measurement of all instruments. This is due primarily to the screw mechanism, and the fact that there are few moving parts. The micrometer discussed is a mechanical unit and requires the operator to read lines and numbers. The newer units are digital and require the operator to read the numbers. To carry the micrometer even further, the micrometer is connected to a computer allowing for not only a larger display of the readings but a record of the readings taken throughout a measuring session. The technician should become familiar with precision measuring devices. Mechanical technology is shrinking and where measurements in the past were made with a scale, a micrometer is needed today and tomorrow.

6-2   Dial Caliper

The dial caliper is an instrument used to measure inside, outside, and depth dimensions (**Figure 6-5**). It is used to measure accuracy from approximately .001 of an inch to .0001 of an inch. It is easy to use, read, and understand.

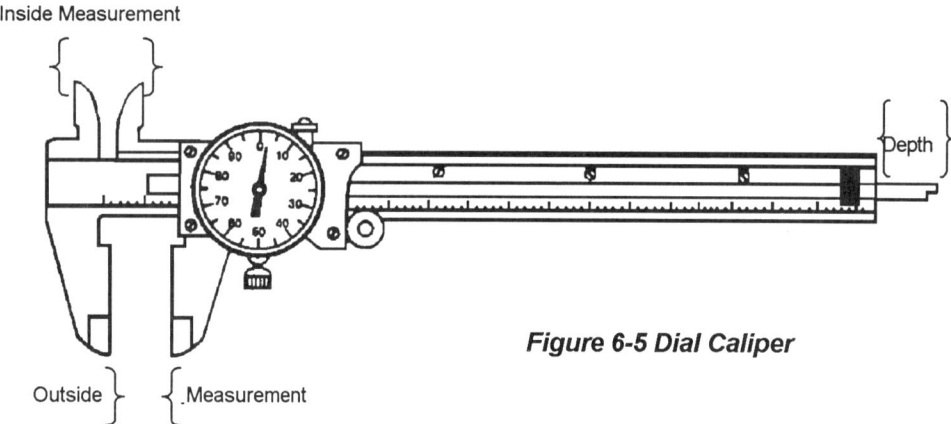

Figure 6-5 Dial Caliper

The dial caliper is an outgrowth of the vernier caliper. However, the beam scale on the dial caliper is graduated into .10 of an inch increments. The caliper dial is either graduated into 100 or 200 divisions. The dial hand is operated by a pinion gear that engages a rack on the caliper beam. On the 100-division dial, the hand makes one complete revolution for each .01 of an inch movement of the sliding jaw along the beam. Therefore, each dial graduation represents 1/100 of .01 of an inch, or .001 of an inch maximum discrimination (precise reading). If readings exceed 1", the readings should be **read** as 1 and however many thousands of an inch. To **write** it as a number; it is written as 1.234 of an inch.

Summary

The dial caliper is directly read which greatly facilitates reading of the instrument. For this reason, the dial caliper has all but replaced its vernier counterpart in many applications. When using the dial caliper, remember the expectation of accuracy in caliper instruments.

6-3   Dial Indicator

The dial indicator is a precision measuring instrument that is used to measure out of round parts. It is able to show a plus or minus movement. It is also used to measure the total movement of a mechanical device. Example: the total movement of a cam, from start to finish. It can also be used to detect a shaft on a motor whose bearings are worn, and the eye cannot detect, but vibration can be felt.

The instrument can also be set to measure parts being manufactured to separate acceptable tolerances from unacceptable tolerances.

The dial indicator is a valuable and useful tool.

It generally takes the form of a spring-loaded spindle that, when depressed, actuates the hand of an indicating dial. Note that the dial face is usually graduated in thousandths of an inch or subdivisions of thousandths. This might lead you to believe that the indicator spindle movement corresponds directly to the amount shown on the indicator face. However, this conclusion is to be arrived at only with the most cautious judgment. **Dial test indicators should not be used to make direct linear measurements.** Dial indicators can be used to make linear measurements, but only if they are specifically designed to do so and under proper conditions.

Dial indicators have discriminations that typically range from .00005 to .001 of an inch. Indicators are equipped with a rotating face or bezel. This feature permits the instrument to be set to zero at any desired place. Many indicators also have a bezel lock. Dial indicators may have removable spindle tips, thus permitting use of different shaped tips as required by the specific application.

All indicators must be mounted solidly if they are to be reliable. Indicators must be clamped or mounted securely when used on a machine tool. A number of mounting devices are in common use. Some of these have magnetic bases that permit an indicator to be attached at any convenient place on the machine tool. This useful magnetic base, shown in **Figure 6-6**, has a provision for turning off the magnet by mechanical means.

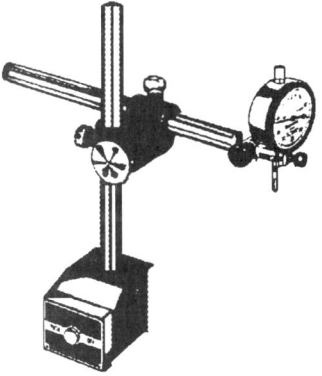

Figure 6-6 Magnetic Base Dial Indicator

CAUTION: Dial indicators are precision instruments and should be treated accordingly. They should NOT be dropped or exposed to severe shocks.

Summary

The dial indicator is a very useful measuring device. It can be used with the machine or device operating. It can be used to measure bi-direction or set for one direction measurement. Since measurements being made are in the thousands of an inch which is the thickness of a human hair. With machine size shrinking, making measurements on such equipment relies on precision measuring instruments, which are becoming more important to technicians. It is therefore important to have a good understanding of the operation of the dial indicator.

6-4   **Measuring Devices Exercises**

Measuring bolts using the dial calipers.

To determine the specifications of a bolt:

> Length of bolt
> Diameter of bolt
> Length of threads
> Number of threads per inch

Follow these procedures using **Figures 6-7** and **6-8** on the next page:

1.   **Length of Bolt** – Using the dial calipers, measure from the end of the bolt to under the head. This is done on all bolts or screws EXCEPT the flat head bolt or screw. Flat head measurements include the head (**Figure 6-7**).

2.   **Diameter of Bolt** – Measure the diameter of the bolt (**Figure 6-7**). After measuring, the decimal diameter must be converted to a stock size. A stock size is one that is readily available. After measuring the bolt, to convert the bolts stock number diameter, use the "BODY SIZE CHART" to locate a bolt dimension that comes closest to the measured dimension. This will serve as the first number of your bolt. The decimal number will be converted to a stock size. Example: .138 would be a No.6, .164 would be a No.8, .190 would be a No.10, etc.

        NOTE: It *MUST* be one of the numbers listed on the chart as those are stock numbers. Stock numbers are those bolts that are easily found. Other sizes are considered specialty sizes.

3.   **Length of Threads** – The length of the threads in SOME cases will be the same as the length of the bolt. If not, use the depth measurement feature on your calipers (**Figure 6-8**).

4.   **Number of Threads Per Inch** – To determine the number of threads per inch, set the dial calipers to .250 of an inch (**Figure 6-8**). Then. using the "depth measure" on the dial calipers, set on the end of the bolt, count the number of full threads, and multiply that number by four (since there are four quarters in an inch). This number is also a stock number and can be found on the chart.

5.   **From the numbers**, the screw size has been determined, whether it is a **6-32, 8-32, 10-24**, etc. Then, from these numbers, and using a drill chart (see Chapter 7), the proper drill for tapping or clearance can be determined.

To determine thread
size (no. threads/inch),
set calipers to .250 (1/4
of an inch)

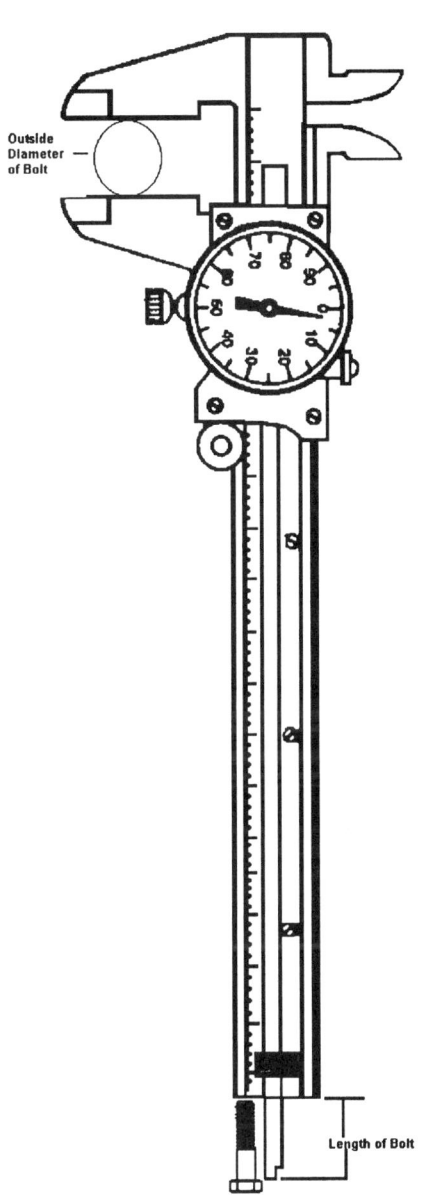

Figure 6-7

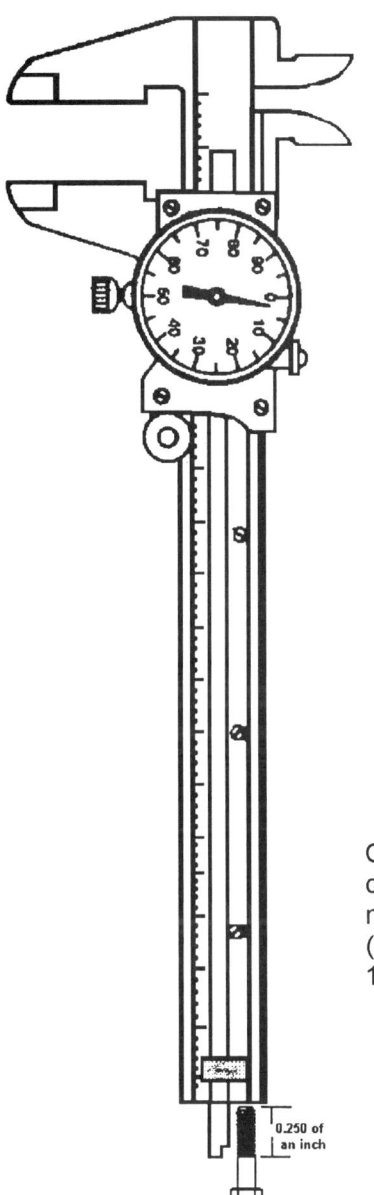

Count the number
of threads and
multiply by 4.
(Note: there are 4
1/4's in an inch)

Figure 6-8

Using the Dial Calipers to Measure a Bolt

Bolt Measurement Worksheet

Using the dial calipers, identify several different types of bolts.

1.   The first type should be a hex head bolt similar to (**Figure 6-9**). In measuring bolts, the head is not included in the measurement.

    a.   **MEASURE** the length of bolt _____

    b.   **MEASURE** the length of threads _____

    c.   **MEASURE** the diameter of bolt _____

    d.   **COUNT** the number of threads per inch _____

    e.   **DETERMINE** the drill needed to *tapped* hole _____

    f.   **DETERMINE** the drill needed to drill a *clearance* hole _____

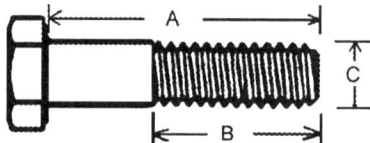

Figure 6-9 Hex Head Bolt

2.   The next type should be a flat head screw (**Figure 6-10**). **Note: the difference in measurement the head is included in the measurement.**

    a.   **MEASURE** the length of hardware_____

    b.   **MEASURE** the length of threads_____

    c.   **MEASURE** the diameter of hardware_____

    d.   **COUNT** the number of threads per inch_____

    e.   **DETERMINE** the drill needed to *tapped* hole_____

    f.   **DETERMINE** the drill needed to drill a *clearance* hole_____

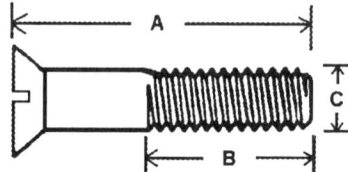

Figure 6-10 Flat Head Screw

Summary

This section covered various types of precision measuring devices: micrometer, dial caliper, and dial indicator. There were practice exercises to familiarize the technician with their proper readings. This section did not cover all precision devices but addressed only three basic types. To some the micrometer is the most difficult to read. Using the exercises will make it easier. Using the dial calipers to determine inside outside and depth measurements can also be useful when duplicating various components. Once the technician understands how to use the measuring devices, with practice reading precision tools will become easier. It should be noted that precision is essential in today's equipment. The three basic measuring tools presented are only a fraction of the precision measuring tools available to the technician. A basic understanding of the tools presented will make reading of other precision tools easier to master.

SECTION II – HAND TOOLS AND HARDWARE

CHAPTER 7 – DRILLING

Introduction   Drilling holes is one of the most basic of machining operations. Metal cutting requires considerable pressure of feed on the cutting edge. A drill press provides the necessary feed pressure either by hand or power drive. The primary use of the drill press is to drill holes. It can be used for other operations such as counter-sinking, counter-boring, spot facing, reaming, and tapping which are processes that modify the drilled hole.

Objectives   After completing this chapter, you should be able to:

♦   Identify different types of drills.

♦   Determine which drill bit should be used to drill a specific type of hole.

♦   Recognize the various parts of a drill press.

♦   Identify the proper lubricant when using power tools or hand tools.

♦   Know safety rules when working with power tools.

7-1   Drilling Tools

Drilling – Using a drill bit to drill a specific size hole for the purpose of connecting two or more pieces of metals together (**Figure 7-1**).

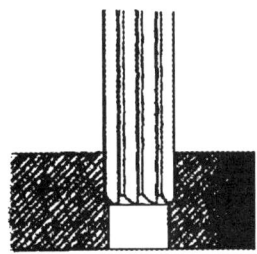

Reaming – A reaming tool is used after drilling a hole. It has sharp straight side edges that, when forced into the pre-drilled hole, produces a clean straight hole in the material (**Figure 7-2**).

Boring – This tool is used to produce a hole using a specially designed cutting point. It can also be used to open a pre-existing hole. The requirements for hole design are specified by product design (**Figure 7-3**).

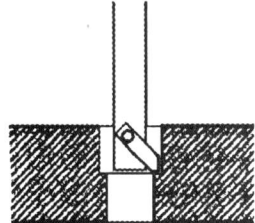

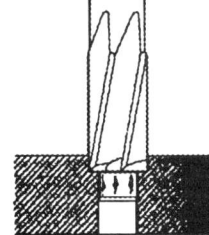

Counter-Boring – This type of drill is used to produce a hole in which the head of the screw fits into the hole with the threads. In this case, alignment is critical (**Figure 7-4**).

Counter-Sinking – This tool produces a "V"-grooved hole in the material which is used for a counter-sunk screw. The head is flush with the surface area (**Figure 7-5**).

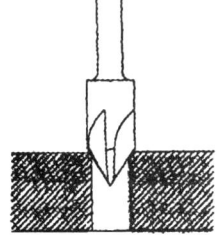

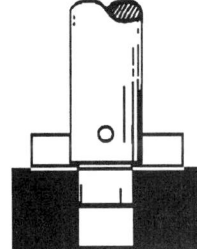

Spotfacing – This tool is used to remove some of the surface area's material to allow the head of the screw to be slightly below the surface material (**Figure 7-6**).

Tapping – This tool produces threads in a properly drilled hole. Note: when using a drill press to tap holes, the drill press must have a reverse, and a pressure relief mechanism to prevent the tap from breaking. The drill press can be used to start a tap. This aids in starting the tapping process in a straight line with the drilled hole (**Figure 7-7**). Tapping will be covered in greater detail in **Chapter 9**.

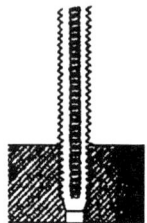

7-2   Drill Presses

There are three basic types of drill presses used for general drilling operations: the **sensitive drill press**, the **upright drilling machine**, and the **radial arm drill press**. Each type has its own characteristics. They all perform the same process: drilling holes. Some have more capabilities than others. The most widely used is the sensitive type drill press.

Sensitive Drill Press – The sensitive drill press, as the name applies, allows the operator to "feel" the cutting action of the drill as the work is fed into it. These machines are either bench or floor mounted. Since these machines are used for light applications only, they usually have a maximum drill size of ½" diameter. Machine capacity is measured by the diameter of work that can be drilled.

The sensitive drill press has four major parts, not including the motor: the head, column, table, and base. **Figure 7-8** labels the parts of a drill press that you should remember. The spindle rotates within the quill, which does not rotate but carries the spindle up and down. The spindle shaft is driven by a **stepped-vee pulley and belt, which requires the operator to change speeds** or by a **variable speed drive** which has a special speed adjustment control.

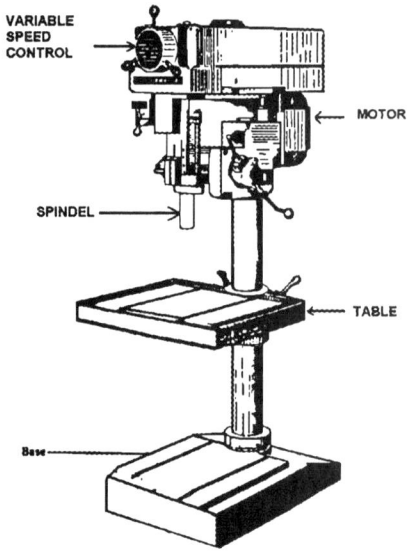

NOTE: THE MOTOR MUST BE RUNNING AND THE SPINDLE TURNING WHEN CHANGING SPEEDS WITH A VARIABLE SPEED DRIVE. ON NON-VARIABLE SPEED DRILL PRESSES, UNPLUG THE DRILL PRESS BEFORE CHANGING BELT SPEED.

Figure 7-8 Drill Press Components

Upright Drill Press

The upright drill press is very similar to the sensitive drill press, but it is made for much heavier work. The drive is more powerful and many types are gear driven, therefore, they are capable of drilling holes up to two inches or more in diameter.

Radial Arm Drill Press

The radial arm drill press is the most versatile drilling machine. Its size is determined by the diameter of the column and the length of the arm measured from the center of the spindle to the outer edge of the column. It is very useful for operations on large castings that are too heavy to be repositioned by the operator for drilling each hole. The work is clamped to the table or base, and the drill can then be positioned where it is needed by swinging the arm and moving the head along the arm. The radial arm drill press is used for drilling small to very large holes and for boring, reaming, counter-boring, and counter-sinking. Like the upright machine, the radial arm drill press has a power feed mechanism and a hand feed lever.

CAUTION: It is important to remember that any drill press is potentially dangerous. If strict safety rules are not used, damage to the machine can be expected.

WARNING: Because of the potential danger involved with any drill press, follow all safety precautions. Serious injury or death could result if these safety rules are not followed. **Always wear goggles!!**

Summary

The drill press has many advantages over hand drills. It can also be a dangerous piece of machinery. Caution and following the basic safety rules can prevent serious harm to the operator. Before operating machinery have a qualified operator explain the proper operating procedures of that machine. DON'T be a statistic, or a casualty. Follow drill press and hand drill safety rules.

7-3   Drill Press and Hand Drill Safety Rules

When operating power tools, it is important to observe basic safety rules. These rules have been tried and tested in the work place. These general safety rules, if followed, will prevent the operator from being injured.

1. Remove neckties and tuck in loose clothing so there is no chance of them becoming entangled with the rotating drill.

2. Check out the machine. Are all the guards in place? Do the switches work? Does the machine operate properly? Are the tools sharpened properly for the material being worked?

3. Clamp the work solidly. DO NOT hold it with your hands. A "merry-go-round" can inflict serious and painful injuries.

4. Wear goggles.

5. Place a piece of wood under the drills being removed from the machine. Small drills are damaged in dropping and the larger tools can injure you in dropping if they fall on your foot.

6. Use sharp tools.

7. Clean chips from the work area with a brush, NOT your hands.

8. Treat cuts and scratches immediately.

9. Always remove the key from the chuck before turning the power on.

10. Let the drill spindle stop of its own accord after the power has been turned off. DO NOT try to stop it with your hands.

11. Wipe up all cutting fluid that spills on the floor.

12. Never clean the tapered opening in the spindle while the machine is operating.

13. After using the drill, wipe the drill clean of chips and cutting fluid with a rag. DO NOT use your hands.

14. Place all oily and dirty waste in a closed container when the job is finished.

The following lubricants are recommended for use when drilling or tapping holes with *power tools and hand tools*. It should be noted that when using **high speed machines** for metal cutting, there are other lubricants that are recommended.

Steel	Cutting Oil
Aluminum	Kerosene
Magnesium	Kerosene
Brass	Kerosene
Copper	Kerosene
Zinc	Kerosene
Cast Iron	Dry
Bakelite	Dry

Summary

Power tool safety is very important to the technician. If the technician does not follow the basic safety rules, injury or death can occur. Safety is everyone's responsibility, not just the operator. If someone is not following the rules, remind them, and encourage them to follow them. It only takes a split second to cause injury but it can take up to a year to recover or recovery may not be possible. REMEMBER: SAFETY-SAFETY-SAFETY – it's your responsibility.

7- 4   Drill Selection Table

TAP DRILLS

American Std. And unified Form Threads Tap Drill Size is approximately 75% Thread

Thread Norm Size	Pitch Series	Drill Size	Drill Decimal
0-80	NF	3/64	.047
1-64	NC	53	.060
1-72	NF	53	.060
2-56	NC	50	.070
2-64	NF	50	.070
3-48	NC	47	.079
3-56	NF	45	.082
4-40	NC-UNC	43	.089
4-48	NF	42	.094
5-40	NC	38	.102
5-44	NF	37	.104
6-32	NC-UNC	36	.107
6-40	NF	33	.113
8-32	NC-UNC	29	.136
8-36	NF	29	.136
10-24	NC-UNC	25	.150
10-32	NF-UNF	21	.159
12-24	NC	16	.177
12-28	NF	14	.182
1/4-20	NC-UNC	7	.201
1/4-28	NF-UNF	3	.213
5/16-18	NC-UNC	F	.257
5/16-24	NF-UNF	I	.272
3/8-16	NC-UNC	5/16	.313
3/8-24	NF-UNF	Q	.332
7/16-14	NC-UNC	U	.368
7/16-20	NF-UNC	25/64	.391

BODY SIZES	COURSE	FINE	
SIZE	THD's Per In	THD's Per In	DIAM
0		80	0.060
1	64	72	0.073
2	56	64	0.086
3	48	56	0.099
4	40	48	0.112
5	40	44	0.125
6	32	40	0.138
8	32	36	0.164
10	24	32	0.190
12	24	28	0.216
1/4	20	28	0.250
5/16	18	24	0.312
3/8	16	24	0.375
7/16	14	20	0.437
1/2	13	20	0.500

FRACTIONAL DRILLS	
FRACTION	DECIMAL
1/64	.015625
1/32	.031250
3/64	.046875
1/16	.062500
5/64	.078125
3/32	.093750
7/64	.109375
1/8	.125000
9/64	.140625
5/32	.156250
11/64	.171875
3/16	.187500
13/64	.203125
7/32	.218750
15/64	.234375
1/4	.250000
17/64	.265625
9/32	.281250
19/64	.296875
5/16	.312500
21/64	.328125
11/32	.342750
23/64	.359375
3/8	.375000
25/64	.390625
13/32	.406250
27/64	.421875
7/16	.437500
29/64	.453125
15/32	.468750
31/64	.484375
1/2	.500000

LETTER DRILLS	
A	.234
B	.238
C	.242
D	.246
E	.250
F	.257
G	.261
H	.268
I	.272
J	.277
K	.281
L	.290
M	.295
N	.302
O	.316
P	.323
Q	.332
R	.339
S	.348
T	.358
U	.368
V	.377
W	.386
X	.397
Y	.404
Z	.413

NUMBER DRILLS	
60	.0400
59	.0410
58	.0420
57	.0430
56	.0465
55	.0520
54	.0550
53	.0595
52	.0635
51	.0670
50	.0700
49	.0730
48	.0760
47	.0785
46	.0810
45	.0820
44	.0860
43	.0890
42	.0935
41	.0960
40	.0980
39	.0995
38	.1015
37	.1040
36	.1065
35	.1100
34	.1110
33	.1130
32	.1160
31	.1200
30	.1285
29	.1360
28	.1405
27	.1440
26	.1470
25	.1495
24	.1520
23	.1540
22	.1570
21	.1590
20	.1610
19	.1660
18	.1695
17	.1730
16	.1770
15	.1800
14	.1820
13	.1850
12	.1890
11	.1910
10	.1935
9	.1960
8	.1990
7	.2010
6	.2040
5	.2055
4	.2090
3	.2130
2	.2210
1	.2280

7-5   Drill Selection Exercise

Upon completion of the drill chart, it will become a handy chart to use to find the proper drill bit to use for drilling a clearance (or body) hold and the proper drill for tapping a hole. The procedure for completing the chart using the Drill Selection Table in 7-4 is listed below.

Screw Size	Tap Drill Decimal	Tap Drill Number	Body Drill Decimal	Body Drill Number
0 - 80				
1 - 64				
1 - 72				
2 - 56				
2 - 64				
3 - 48				
3 - 56				
4 - 40				
4 - 48				
5 - 40				
5 - 44				
6 - 32				
6 - 40				
8 - 32				
8 - 36				
10 - 24				
10 - 32				
12 - 24				
12 - 28				
1/4 - 20				
1/4 - 28				
5/16 - 18				
5/16 - 24				
3/8 - 16				
3/8 - 24				

How to use the drill selection tables in 7-4 to complete the drill chart below.

1. Locate the Thread Norm Size for the screw. Example: 0-80 requires a hole diameter of 0.047. This number should be placed in the column marked Tap Drill Decimal.

2. Next, the size 3/64 drill bit should be placed in the column marked Tap Drill Number.

3. This process or procedure should be repeated for each of the sizes noted for the tap column.

4. Next, use the Body Sizes chart to locate the body size of each of the basic screws listed. Example: Body Size for a 0-80 is 0.060. That number should be placed in the column labeled Body Drill Decimal.

5. The next step requires more searching. A 0-80 screw body is a maximum of 0.060. A drill bit that is exactly 0.060 would be ideal. The closest is a 53 number drill found on the Number Drills chart and its diameter is 0.0595. If there is not one listed at exactly the desired size, then find one that is one to two thousandths larger.

    NOTE: The screw sizes and drills listed are stock sizes and will vary a few thousandths of an inch.

6. Fill in the rest of the two columns using the same procedure.

When finished, a quick reference drill chart for stock hardware will save time looking up drills needed for drilling a body hole or a clearance hole. When specified by the manufacturer, use those specific drills.

SECTION II – HAND TOOLS AND HARDWARE

CHAPTER 8 – METAL CUTTING TOOLS

Introduction The hacksaw is a versatile material cutting tool. It can be used to cut any type of soft metal and some hard metals. The hacksaw is used to cut small parts and trim other materials. It is portable and can accommodate different types of blades for different materials. It is important to know the different blades and procedures needed to make a clean cut.

Objectives After completing this chapter, you should be able to:

  ♦ Identify different blades used to cut high speed steels or soft metals.

  ♦ Identify speeds to cut materials.

  ♦ Identify the cause of problems when the hacksaw will not cut properly.

  ♦ Determine the number of teeth needed on the cutting surfaces of different materials.

8-1   Hacksaws

The Proper Use of a Hacksaw

The parts of a hacksaw are shown in **Figure 8-1**.

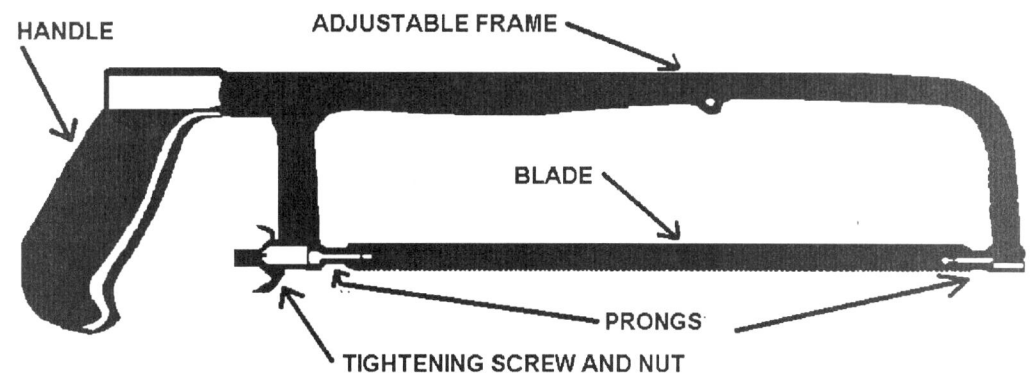

HANDLE   ADJUSTABLE FRAME

BLADE

PRONGS

TIGHTENING SCREW AND NUT

Frames are either the solid or adjustable type. The solid frame can only be used with one length of saw blade. The adjustable frame can be used with hacksaw blades that are 8 to 12 inches long.

The blade can be mounted to cut in line with the frame or at a right angle to the frame. This is done by turning the blade at right angles to the frame. This will allow you to continue a cut that is deeper than the capacity of the frame. If the blade is left in line with the frame, the frame will eventually hit the work piece and limit the depth of the cut. Most hacksaw blades are made from high speed steel, and in lengths of 8, 10, and 12 inches. Blade length is the distance between the centers of the holes at each end. Hand hacksaw blades are generally ½" wide and .025" thick. The kerf or cut produced by the hacksaw is wider than the .025" thickness of the blade because of the teeth (**Figure 8-2**).

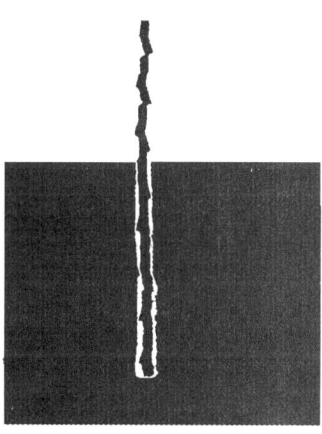

The **set** refers to the bending of teeth outward from the blade itself. Two kinds of set are found on the hand hacksaw blades. The first is the straight or alternate set (**Figure 8-3**) where one tooth is bent to the right and the next tooth to the left for the length of the blade. The second kind of set is the wavy set in which a number of teeth are gradually bent to the right and then to the left (**Figure 8-4**). A wavy set is found on most fine tooth hacksaw blades.

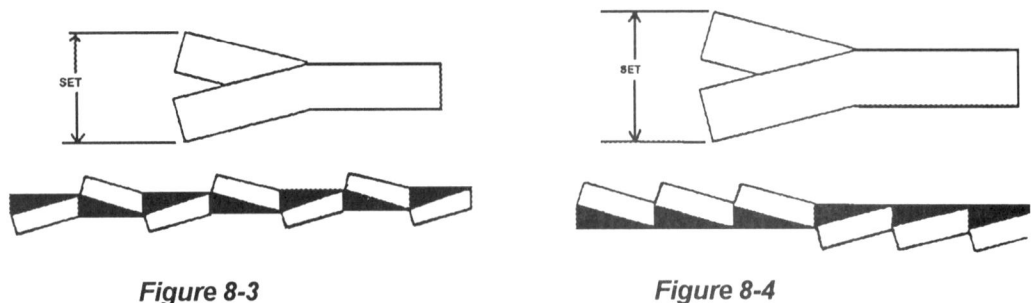

Figure 8-3   *Figure 8-4*

The spacing of the teeth on a hacksaw blade is called the pitch and is expressed in teeth per inch of length (**Figure 8-5**). Standard pitches are 14, 18, 24, and 32 teeth per inch, with the 18-pitch blade used as a general purpose blade.

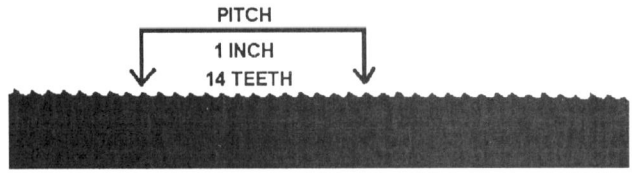

Figure 8-5

Note: A new blade must be started on the opposite side of the work, not in the same kerf as the old blade.

The hardness and size or **thickness** of a work piece determines to a great extent which pitch blade is used. As a rule, you should use a coarse tooth blade on soft materials to have sufficient clearance for the chips and a fine tooth blade on harder materials, but you also should have at least three teeth cutting at any time, which may require a fine tooth blade on soft materials with thin cross sections.

Hand hacksaw blades fall into two categories: soft-backed or flexible blades and all-hard blades. On the flexible blades only the teeth are hardened, the back being tough and flexible. The flexible blade is less likely to break when used in places difficult to get at, such as in cutting off bolts on machinery. The all-hard blade is, as the name implies, hard and very brittle and should be used only where the work piece can be rigidly supported, as in a vise. On an all-hard blade, even a slight twisting motion may break the blade. All-hard blades, in the hands of a skilled person, will cut true straight lines and give long service.

The blades are mounted in the frame with the teeth pointing away from the handle so that the hacksaw cuts only on the forward stroke. No cutting pressure should be applied to the blade on the return stroke as this tends to dull the teeth. The sawing speed of a hacksaw should be 40 to 60 strokes per minute. To get the maximum performance from a blade, make long, slow, and steady strokes using the full length of the blade. Sufficient pressure should be maintained on the forward stroke to keep the teeth cutting. Teeth on a saw blade will dull rapidly if too little or too much pressure is put on the saw. The teeth will dull also if too fast a cutting stroke is used; a speed in excess of 60 strokes a minute will dull the blade because friction will overheat the teeth.

The saw blade may break if it is too loose in the frame or if the work piece slips in the vise while sawing. Too much pressure may also cause the blade to break. A badly worn blade where the set has been worn down will cut a too narrow kerf. This will cause binding and perhaps breakage of the blade. When this happens and a new blade is used to finish the cut, turn the work piece over and start with the new blade from the opposite side and make a cut to meet the first one (**Figure 8-6**). The set on the new blade is wider than the old kerf. Forcing the new blade into an old cut will immediately ruin it by wearing the set down.

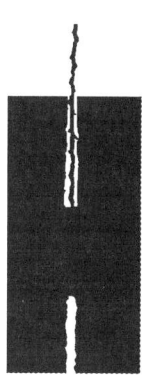

A cut on a work piece should be started with only light cutting pressure, with the thumb or fingers on one hand acting as a guide for the blade. Sometimes it helps to start a blade after a small v-notch is filed in the work piece. When a work piece is supported in a vise, make sure that the cutting is done close to the vise jaws for a rigid setup free of chatter *(**Figure 8-7***). Work should be positioned in a vise so that the saw cut is vertical. This makes it easier for the saw to follow a straight line. At the end of a saw cut, just before the pieces are completely parted, reduce the cutting pressure or you may be caught off balance when the pieces come apart and cut your hands on the sharp edges of the work piece.

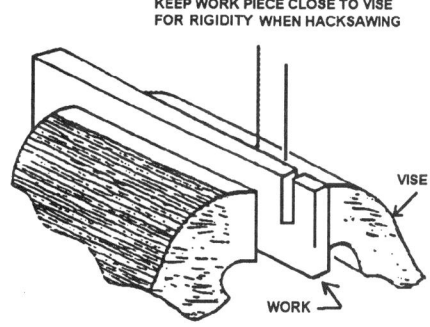

KEEP WORK PIECE CLOSE TO VISE
FOR RIGIDITY WHEN HACKSAWING

VISE

WORK

Figure 8-7. The work piece is cut close to the vise to avoid vibration and chatter.

To saw thin material, sandwich it between two pieces of wood for a straight cut. Avoid bending the saw blades because they are likely to break and, when they do, they usually shatter in all directions and could injure you or others nearby (see **Figure 8-8**). This figure shows the proper way to select a blade for cutting material.

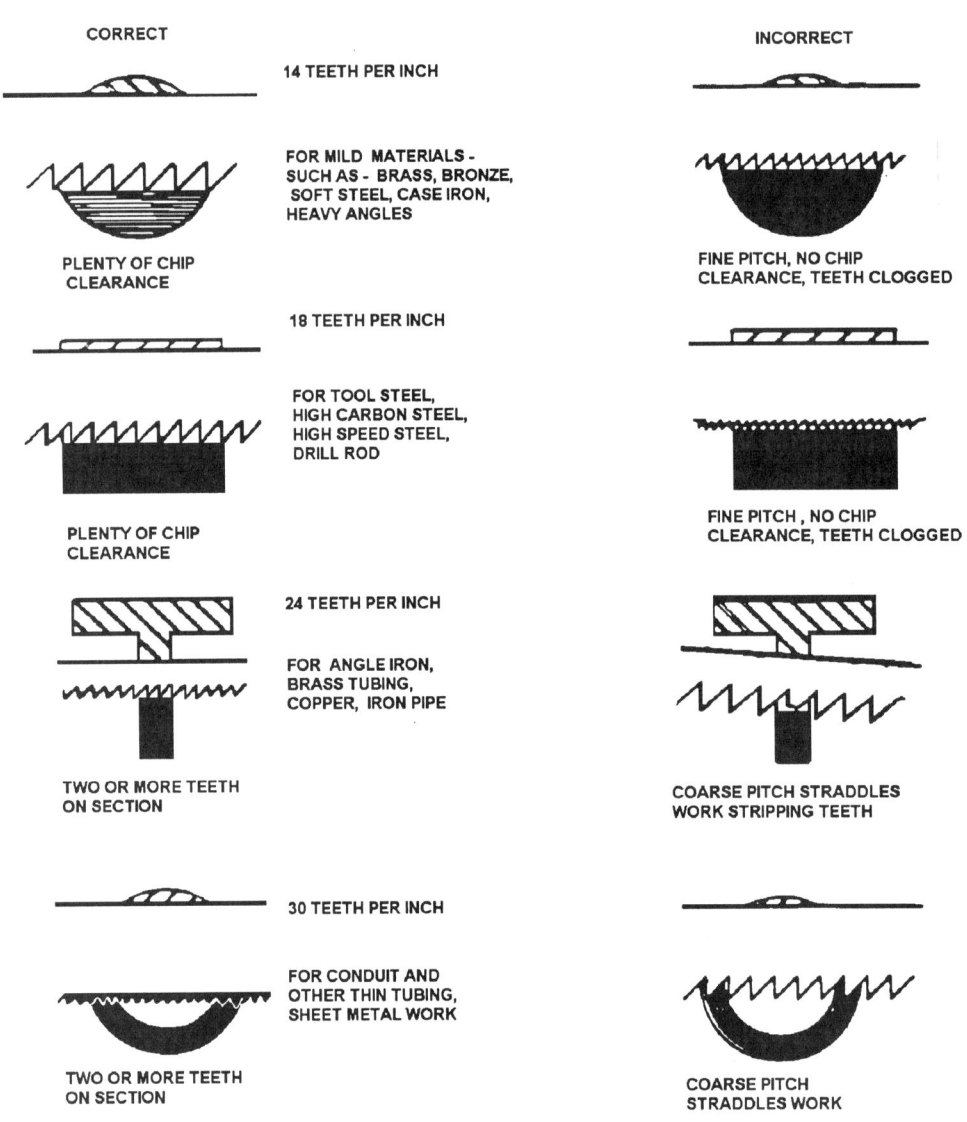

CORRECT

14 TEETH PER INCH

FOR MILD MATERIALS - SUCH AS - BRASS, BRONZE, SOFT STEEL, CASE IRON, HEAVY ANGLES

PLENTY OF CHIP CLEARANCE

18 TEETH PER INCH

FOR TOOL STEEL, HIGH CARBON STEEL, HIGH SPEED STEEL, DRILL ROD

PLENTY OF CHIP CLEARANCE

24 TEETH PER INCH

FOR ANGLE IRON, BRASS TUBING, COPPER, IRON PIPE

TWO OR MORE TEETH ON SECTION

30 TEETH PER INCH

FOR CONDUIT AND OTHER THIN TUBING, SHEET METAL WORK

TWO OR MORE TEETH ON SECTION

INCORRECT

FINE PITCH, NO CHIP CLEARANCE, TEETH CLOGGED

FINE PITCH, NO CHIP CLEARANCE, TEETH CLOGGED

COARSE PITCH STRADDLES WORK STRIPPING TEETH

COARSE PITCH STRADDLES WORK

Figure 8-8

Summary

There is a proper way to cut materials and an improper way. Improper ways result in damage to both the material and the operator. Select the proper blade, secure the material in a vise, then apply the appropriate pressure to the saw. When cutting is finished, the material will have a clean cut. In addition, when finished, remove any sharp edges from the material. Try to keep the cutting straight, as this will prevent excessive filing to correct improper cuts. When unsure of your cutting, practice with a piece of scrap material before cutting the important piece. Remember, **SAFETY** is a must when working with material cutting equipment.

SECTION II – HAND TOOLS AND HARDWARE

CHAPTER 9 – TAPS AND TAPPING

Introduction The mass production of consumer goods today depends to a large extent on the efficient and secure assembly of parts using threaded fasteners. It takes skill to produce usable tapped holes. A craftsman in the metal trades as well as technicians modifying existing products or repairing damaged products, must have an understanding of the factors that effect the tapping of a hole, such as work material and its cutting speed, the proper coolant, and the size and condition of the hole. A good technician can analyze a tapping operation, determine whether or not it is satisfactory, and usually find a solution.

Most internal threads produced today are made with taps. These taps are available in a variety of styles; each one designed to perform a specific type of tapping operation efficiently. This Chapter will help you identify and select taps for threading operations.

Objectives After completing this chapter, you should be able to:

♦ Identify the parts of a tap.

♦ Identify the three basic taps.

♦ Determine which tap is used to tap various types of holes.

♦ Identify different types of taps.

♦ Determine which type of lubricant to use when high speed tapping or hand tapping is required.

♦ Properly tap a hole using hand-tapping procedures.

9-1   Identifying Common Tap Features

Taps are used to cut internal threads in holes. This process is called tapping. Tap features are shown in **Figures 9-1** and **9-2**. The active cutting part of the tap is the chamfer, which is produced by grinding away the tooth formed at an angle, with relief back of the cutting edge, so the cutting action is distributed progressively over a number of teeth. The fluted portion of the tap provides space for chips to accumulate and for the passage of cutting fluids. Two, three, and four flute taps are common.

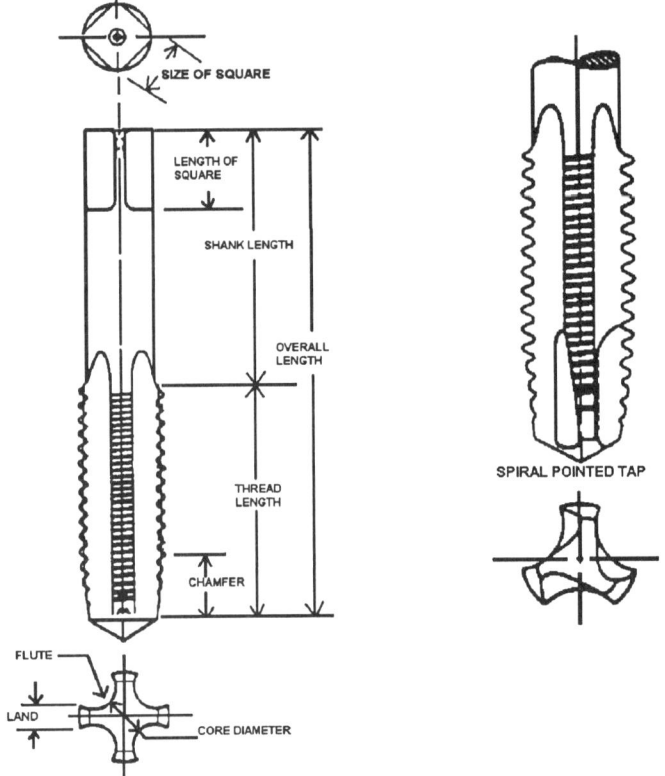

Figure 9-1 Tap Features

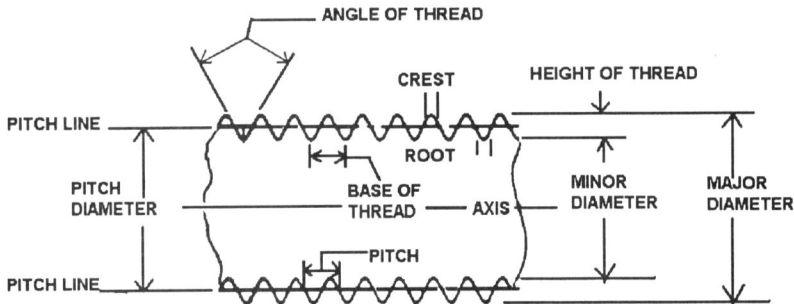

Figure 9-2 Detailed Tap Features

Taps are made from either high carbon steel or high-speed steel, which has a hardness of Rockwell C63. (For further information on hardness of metals, consult the Machinery Handbook.) High-speed steel taps are far more common in manufacturing plants than carbon steel taps. High-speed steel taps typically are ground after heat treatment to ensure accurate thread geometry.

Another identifying characteristic of taps is the amount of chamfer at the cutting end of the tap. A set consists of three taps – **starter, finishing**, and **bottoming** – which are identical except for the numbers of chamfered threads. The **starter** tap is useful in starting a tapped thread square with the part. The most commonly used tap, in hand and machine tapping, is a **finishing** tap. **Bottoming** taps are used to extend almost to the bottom of a blind hole. A blind hole is one that is not drilled clear through a part.

Figure 9-3 shows the identifying markings of a tap, the normal size, the number of threads per inch, whether it is a NC or NF thread series. G is the symbol for ground taps; H3 identifies the tolerance range of the tap, HS means high-speed steel, and LH for left handed markings.

Figure 9-3 Identifying Markings on a Tap

9-2   Other Kinds and Uses of Taps

The most commonly used tap is the hand tap **(Figure 9-4)**. The hand tap is manufactured to produce threads in both machine screw sizes and fraction sizes. Hand taps were originally intended for hand tapping of threads, but are now commonly used in machine production jobs as well.

Figure 9-4 Hand Tap

Spiral fluted taps are made with helical instead of straight flutes **(Figure 9-5)** which draw the chips out of the hole. This kind of tap is also used when tapping a hole that has a keyway or spline as the helical lands of the tap will bridge the interruptions. Spiral fluted taps are recommended for tapping deep blind holes in ductile materials such as aluminum, magnesium, brass, copper, and die-cast metals. Fast spiral flute taps are similar to regular spiral fluted taps, but the faster spiral fluted increases the chip lifting action and permits the spanning of comparably wider spaces.

Figure 9-5 Spiral Fluted Tap

A **pulley tap** (**Figure 9-6**) is used to tap set screws and oil cup holes in the hub of pulleys. The long shank also permits tapping in places that might be inaccessible for regular hand taps. When used for tapping pulleys, these taps are inserted through holes in rims which are slightly larger than the shanks of the taps. These holes serve to guide the taps and assure proper alignment with the holes to be tapped.

Figure 9-6 Pulley Tap

9-4   Reducing Friction in Tapping

Steps may be taken to reduce friction and increase tap life. Surface treatment of taps is often an answer if poor thread forming or tap breakage is caused by chips adhering or welding to the cutting faces – then a tapping lubricant should be used. Different metals require different lubricants. Consult a drilling lubrication chart for proper lubricants for the tapping holes.

9-5   Tapping

Taps are used to cut internal threads in holes. The actual cutting process is called tapping and can be performed by hand or machine. A tap wrench (**Figure 9-7**) or a T-handle wrench (**Figure 9-8**) attached to the tap is used to provide driving torque while hand tapping. To obtain a greater accuracy in hand tapping, a hand tapper is used. This fixture acts as a guide for the tap to ensure that it stays in alignment and cuts concentric threads.

Figure 9-7 Tap Wrench   *Figure 9-8 T-Handle Tap Wrench*

Holes can also be tapped in a drill press that has a spindle reverse switch, which is often foot operated for convenience. Drill presses without reversing switches can be used for tapping with a tapping attachment. Some of these tapping attachments have an internal friction clutch where downward pressure on the tap turns the tap forward and feeds it into the work. Releasing downward pressure will automatically reverse the tap and back it out of the work piece. Some tapping attachments have lead screws that provide tap feed rates equal to the lead of the tap. Most of these attachments also have an adjustment to limit the torque to match the size of the tap, which eliminates most tap breakage.

9-6 Threaded Percentage and Hole Strength

The strength of a tapped hole depends largely on the work piece material, the percentage of full thread used, and the length of the thread. The designer usually selects the work piece material. The percentage of thread produced is dependent on the diameter of the drilled hole. Tap drill charts generally give tap drill sizes to produce 75 percent thread.

In some difficult-to-machine materials such as titanium alloys, high tensile steels, and some stainless steels, 50 to 60 percent thread depths will give sufficient strength to the tapped hole. Threaded assemblies are usually designed so that the bolt breaks before the threaded hole strips. Common practice is to have a bolt engage a tapped hole by 1 to 1½ times its diameter.

9-7 Drilling the Correct Hole Size

The condition of the drilled hole affects the quality of the thread produced, as an out-of-round hole leads to out-of-round threads. Bell-mouthed holes will produce bell-mouthed threads. When an exact hole size is needed, the hole should be reamed before tapping. This is especially important for large diameter taps and when fine pitch threads are used. The size of the hole to be drilled is usually obtained from **TAP DRILL CHARTS** (**Figure 7-4**), which usually show a 75 percent thread depth.

Lubrication is one of the most important factors in a tapping operation. Cutting fluids used when tapping serve as coolants, but are more important as lubricants. It is important to select the correct lubricant because the use of the wrong lubricant may give results that are worse than if no lubrication were used. For lubricants to be effective, they should be applied in sufficient quantity to the actual cutting area in the hole.

MATERIALS	SPEEDS in ft/min	LUBRICANT
Aluminum	90 - 100	Kerosene & light oil
Brass	90 - 100	Kerosene & light oil
Cast Iron	70 - 80	Dry or soluble oil
Magnesium	20 - 50	Light based oil
Phosphor bronze	30 - 60	Light based oil
Plastics	50 - 70	Dry or jet air
Steels		
low carbon	40 - 60	Sulfur based oil
high carbon	25 - 35	Sulfur based oil
Molybdenum	10 - 35	Sulfur based oil
Stainless	10 - 35	Sulfur based oil

Table 1. Recommended Cutting Speeds and Lubricants for Machine Tapping

9-8   Solving Tap Problems

Occasionally it becomes necessary to remove a broken tap from a hole. If a part of the broken tap extends out of the work piece, removal is relatively easy with a pair of pliers. If the tap breaks flush with or below the surface of the work piece, a tap extractor can be used. Before trying to remove a broken tap, the chips in the flutes should be removed. A jet of compressed air or fluids can do for this.

NOTE: ALWAYS STAND ASIDE WHEN CLEANING OUT HOLES WITH COMPRESSED AIR AS CHIPS AND PARTICLES TEND TO FLY OUT AT HIGH VELOCITY.

When the chips are packed so tightly in the flutes or the tap is jammed in the work so that a tap extractor cannot be used, the tap may be broken up with a pin punch and removed piece by piece. If the tap is made from common carbon steel and cannot be pin punched, the tap can be annealed (softening the material) so it becomes possible to drill out.

9-9   Tapping Procedure, Hand Tapping

Determine the **size of the thread** to be *tapped* and **select a tap**. Select the proper **tap drills** from the Tap Drill Chart (Figure 7-4). A taper tap should be selected for hand tapping; or if a drill press or tapping machine is to be used for alignment, use a plug tap. Drill the hole using the recommended coolant. Check the hole size.

Countersink the hole entrance to a diameter slightly larger than the major diameter of the threads. If a countersink is not available, then use a large size drill bit to chamfer the hole. This allows the tap to be started more easily, and it protects the start threads from damage.

Mount the work piece in a bench vise so that the hole is in a vertical position. Tighten the tap in the tap wrench. Cup your hand over the center of the wrench and place the tap in the hole in a vertical position. Start the tap by turning two or three turns in a clockwise direction for right hand thread. At the same time, keep a steady pressure downward on the tap.

After the tap is started for several turns, remove the tap wrench without disturbing the tap. Place the blade of a square against the solid shank of the tap to check for squareness. Check from two positions 90° apart. If the tap is not square with the work, it will ruin the thread and possibly break in the hole if you continue tapping.

Back the Tap Out of the Hole and Restart

Use the correct cutting oil on the tap when cutting threads. Turn the tap clockwise one-quarter to one-half turn and then turn it back a three-quarter turn to break the chip. This is done with a steady motion to avoid breaking the tap. **This is commonly referred to as the backoff procedure when tapping.**

When tapping a blind hole, use the taps in the order of starting, plug, and then bottoming. Remove the chips from the hole before using the bottoming tap and be careful not to hit the bottom of the hole with the tap.

A 60° point center is chucked in a drill press to align a tap squarely with the previously drilled hole. Only very slight follow-up pressure should be applied to the tap. Too much downward pressure will cut a loose, oversize thread.

The technician should practice tapping holes on scrap materials to perfect their skills. The first starter materials should be ¼" aluminum stock. Then move to mild steels for practice. It is important to practice on scrap. Do not wait until the equipment needs tapping to practice.

Summary

Learning how to tap a hole to perform modifications or make repairs is a necessary skill in today's market. The technician is expected to perform many new tasks that were done by specialists in the past. They are, in most cases, the only one available to perform such tasks. Therefore, the technician must be prepared for any task they are called upon to do.

SECTION II – HAND TOOLS AND HARDWARE

CHAPTER 10 – THREAD CUTTING DIES AND THEIR USES

Introduction When working with different electromechanical devices, there are times when a stud becomes damaged and replacement is too costly. With a die, most damaged studs can be repaired. There are also times when a piece of round stock is needed to be threaded by hand. This chapter will identify different parts of the die and tools used with the dies. A basic understanding of how to use those dies can save time and money when making repairs on damaged round stock items.

Objectives After completing this chapter, you should be able to:

 ♦ Identify the various parts of dies and die holders.

 ♦ Identify different types of dies.

 ♦ Thread a piece of round stock by following a thread cutting procedure.

10-1   Die Definition

A die is used to cut external threads on the surface of a bolt or rod. Many machine parts and mechanical assemblies are held together with threaded fasteners, most of which are mass produced. Occasionally, however, a technician has to make a bolt or extend the threads on a bolt using a die. The purpose of this section is to introduce to you to some dies and their hand threading uses.

Dies are used to cut external thread on round materials. Some dies are made from carbon steel, but most are made from high-speed steel. Dies are identified by the markings on their face as to the size of thread, number of threads per inch, and form of thread, such as NC, UNF, or other standard designators (see **Figure 10-1**).

Figure 10-1 Die

Common Types of Hand Threading Dies – The die shown in **Figure 10-1** is an example of a round split adjustable die, also called a button die. These dies are made in all standardized thread sizes up to 1½" thread diameters and ½" pipe threads. The outside diameters of these dies vary from 1 to 3 thousandths of an inch.

Adjustments on these dies are made by turning a fine pitch screw that forces the sides of the die apart or allows them to spring together. The range of adjustment of round split adjustable dies is very small, allowing only for loose or tight fit on a threaded part. Adjustments made to obtain threads several thousandths of an inch oversize will result in poor die performance because the heel of the cutting edge will drag on the threads. Excessive expansion may cause the die to break.

Some round split adjustable dies do not have the built-in adjusting screw. Adjustments are then made with the three screws in the die stock (**Figure 10-2**).

Figure 10-2 Die Holder for Adjustable Round Split Dies

Two of these screws on opposite sides of the die stock hold the die in the die stock and also provide opening pressure. The third screw engages the split in the die and provides opening pressure. These dies are used in a die stock for hand threading or in a machine holder for machine threading.

Another type of threading die is the TWO PIECE DIE, whose halves are called blanks. These blanks are assembled in a COLLET consisting of a CAP and GUIDE. The normal position of the blanks in the collet is indicated by witness marks (**Figure 10-3**). The adjusting screws allow for precise control of the cut thread size.

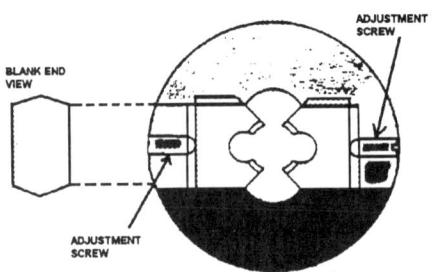

Figure 10-3 Two Piece Die

The blanks are inserted in the cap with the tapered threads toward the guide. Each of the two die halves is stamped with a serial number. Make sure the halves are stamped with a serial number and that each has the same serial number. The guide used in the collet serves as an aid in starting and holding the dies square with the work being threaded. Each thread size uses a guide of the same nominal or indicated size. Collets are held securely in the die stocks **(Figure 10-4)** by a knurled set screw that seats in a dimple in the cap.

Figure 10-4 Collet

Hexagon rethreading dies **(Figure 10-5)** are used to recut slightly damaged or rusty threads. Rethreading dies are driven with a wrench large enough to fit the die. Solid square dies **(Figure 10-6)** have the same uses and limitations as hexagon rethreading dies. All of the previously discussed die types are available in bolt and pipe sizes. Square dies are used to cut new threads and have sufficient chip clearance for this purpose.

Figure 10-5 Hex Die   *Figure 10-6 Square Die*

10-2 Hand Threading Procedures

Threading of a rod should always be started with the leading or throat side of the die. This side is identified by the chamfer on the first two or three threads and also by the size markings. The chamfer distributes the cutting load over a number of threads, which produces better threads and less chance of chipping the cutting edges of the die. Cutting oil and other threading fluids are very important to obtaining quality threads and maintaining long die life. Once a cut is started with a die, it will tend to follow its own lead, but uneven pressure on the die stock will make the die cut variable helix angle or "drunken" threads.

Threads cut by hand often show a considerable accumulated lead error. The lead of a screw thread is the distance a nut moves on the screw if it is turned one full revolution. The problem is caused by the dies being relatively thin when compared to the diameter of thread that they cut. Only a few threads in the die can act as a guide on the already cut threads. This error usually does not cause problems when standard or thin nuts are used on the threaded part. However, when an item with a long internal thread is assembled with a threaded rod, it usually gets tight and then locks, not because the thread depth is insufficient, but because there is a lead error. This error can be as much as one-fourth of a thread in one inch of length.

The outside diameter of the material to be threaded should not be over the nominal size of the thread and preferably a few thousandths of an inch (.002 to .005 inch) undersize. After a few full threads are cut, the die should be removed so that the thread can be tested with a nut or thread ring gauge. A thread ring gauge set usually consists of two gauges, a go and a no-go gauge. As the names imply, a go gauge should screw on the thread, while the no-go gauge will not go more than 1½ turns on a thread of the correct size. Do not assume that the die will cut the correct size thread; always check by gauging or assembling. Adjustable dies should be spread open for the first cut and set progressively smaller for each pass after checking the thread size.

It is very important that a die is started squarely on the rod to be threaded. A lathe can be used as a fixture for cutting threads with a die. The rod is fastened in a lathe chuck for rotation, while the die is held square because it is supported by the face of the tailstock spindle. The carriage or the compound rest prevents the die stock from turning while the chuck is rotated by hand. As the die advances, the tailstock spindle is also advanced to stay in contact with the die. Do not force the die with the tailstock spindle, or a loose thread may result. A die may be used to finish to size a long thread that has been rough threaded on the lathe.

Occasionally, a die is used to extend the thread on a bolt. Make certain that the bolt is not hardened or the die will be ruined. To cut full useable threads close to a shoulder, first cut the threads normally until the die touches the shoulder, then reverse the die and use the unchamfered side to finish the last few threads.

It is always a good practice to chamfer the end of a work piece before starting a die (**Figure 10-7**). The chamfer on the end of a rod can be made by grinding on a pedestal grinder or by filing with a lathe. This will help in starting the cut and it will also leave a finished thread end. While cutting threads with a hand die, the die rotation should be reversed after each full turn forward to break the chips into short pieces that will fall out of the die. Chips jammed in the clearance holes will tear the thread.

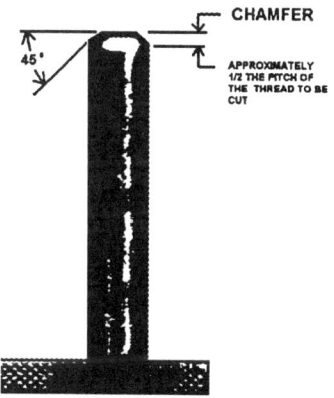

Figure 10-7 Round Stock with Chamfer

10-3   Threading Procedures, Threading Dies

1.   Select the work piece to be threaded and measure its diameter. Then chamfer the end. This may be done on a grinder or with a file. The chamfer should be at least as deep as the thread to be cut.

2.   Select the correct die and mount it in a die stock. Note: Be sure to mount the die so that the start side is toward the work piece. Chamfer side down.

3.   Mount the work piece in a bench vise. Short work pieces are mounted vertically and the long pieces usually are held horizontally.

4.   To start the thread, place the die over the work piece. Holding the die stock with one hand, apply downward pressure and turn the die.

5.   When the cut has started, apply cutting oil to the work piece and die and start turning the die with both hands. After each complete revolution forward, reverse the die one-half turn to break the chips.

6.   Check to see that the thread is started square using a machinists square. Corrections can be made by applying a slight downward pressure on the high side while turning.

7.   When several turns of the thread have been completed, check the fit of the thread with a nut, thread ring gauge, thread micrometer, or the mating part. If the thread fit is incorrect, adjust the die with the adjustment screws and take another cut with the adjusted die. Continue making adjustments until the proper fit is achieved.

8.   Continue threading to the required thread length. To cut threads close to a shoulder, invert the die after the normal threading operation and cut the last two or three threads with the side of the die that has no chamfer.

Summary

Dies are useful tools. They are used to either cut new threads on round stock or chase threads that have been damaged. There are a wide variety of dies to choose from – round, square, or hex.

When using dies, be sure to use the proper handle with the die. This will prevent damage to the die. The technician should know how to use dies. The proper use of dies will produce a quality repair. Keep dies clean and properly stored and they will last many years. A good set of dies should be included in a service technician's toolbox.

SECTION III – MECHANICAL TECHNOLOGY

CHAPTER 11 – MECHANICAL DEVICES

Introduction   This chapter includes the study of some basic mechanical parts used in machines. There are many different devices that are used to physically make things happen in VCR's, printers, fax machines, and copiers. Some of those parts are cams, cam followers, gears, sprockets, idler sprockets, chain drives, pulleys, belts, clutches, thrust bearings, and bushings. Each device has a specific purpose in a machine. The name of each device and the function it performs is discussed.

Objectives   After completing this chapter, you should be able to:

- ◆ Identify basic cams, their parts, and movements.

- ◆ Name the three basic types of gears.

- ◆ Distinguish between the pinion and the gear.

- ◆ Determine which gears use angular transmitting motion.

- ◆ Recognize the four basic sprockets used with roller chains.

- ◆ Determine the location and purpose of an idler sprocket.

- ◆ Distinguish between pulleys, belts, and idler pulleys.

- ◆ Distinguish between a thrust bearing and a bushing.

- ◆ Distinguish between a friction and a positive contact clutch.

- ◆ List the applications of basic bushings.

11-1 Cams and Cam Followers

Cam – A cam is a machine element designed to generate a desired motion in a follower by means of direct contact. Cams are mounted on rotating shafts and convert rotary motion to oscillations. They can also convert motion from one form to another (rotary to reciprocating) (see **Figure 11-1**).

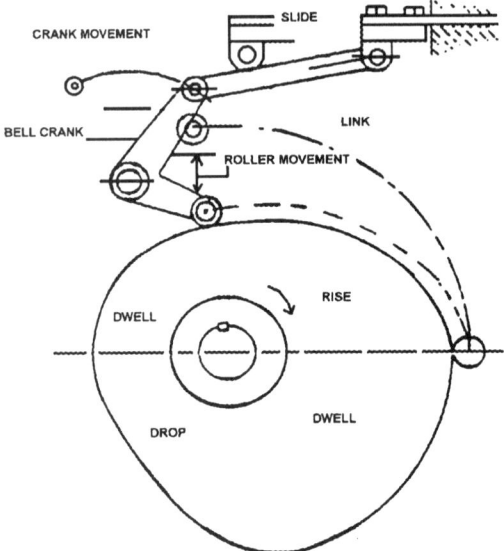

Figure 11-1 Action of the Cam Follower

Cams are machined parts that have irregular curved outlines, a curved outline or a curved groove (see **Figure 11-2**). When the cam moves, it gives a specific motion to another machine part that is called the follower. The cam and the follower together make up the cam mechanism. The cam drives the follower. Cams in general are divided into two classes: uniform motion cams and accelerated motion cams.

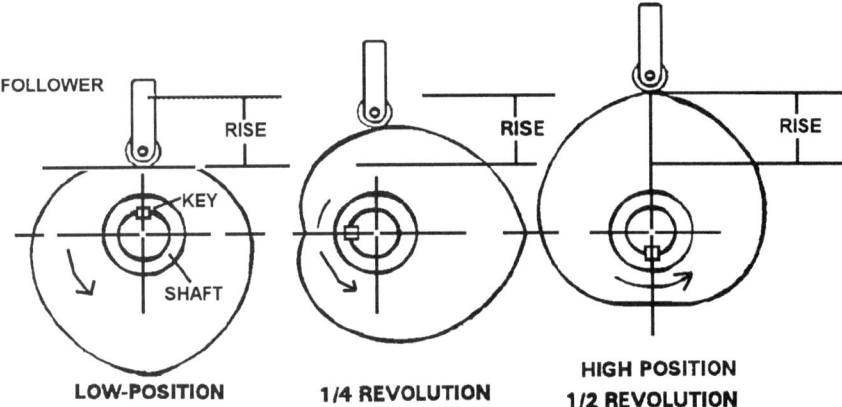

Figure 11-2 Cam Position

The uniform motion moves the follower at the same rate of speed from the beginning to the end of the stroke. However, on start-up and shut-down, the motion is abrupt.

The uniformly accelerated motion cam like **Figure11-2** is suitable for moderate speeds, but has the disadvantage of sudden changes in the beginning or middle of the stroke.

Cam Follower – Cam followers are considered part of the cam and activate electrical or mechanical devices. There are three basic types of followers (see **Figures 3, 4,** and **5**) with four types of cam's using three basic cam followers; **Roller**, **Flat**, and **Point**. Flat cam followers require a spring to maintain pressure on the cam through the cams complete revolution. Another example of a cam and follower is shown in **Figure 11-6.** This shows a Cone Cam and with a roller follower.

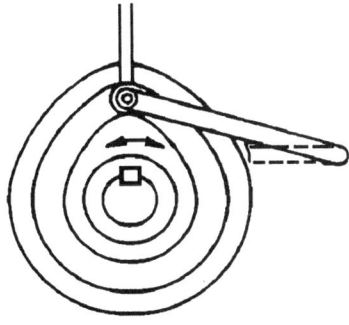

Figure 11-3 Roller Cam Follower

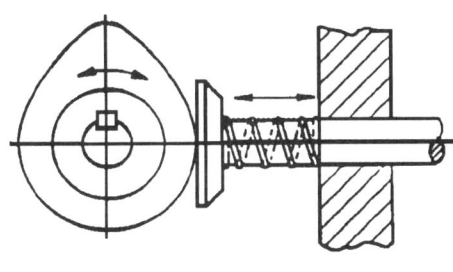

Figure 11-4 Flat Cam Follower

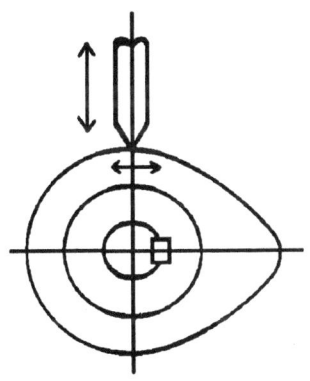

Figure 11-5 Point Cam Follower

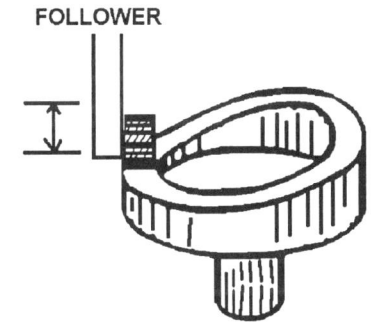

Figure 11-6 Cone Cams and Roller Follower

The flat cam follower has three types of contact points: the roller for high speeds, the point, and flat surface followers for slow speed applications. They are usually made with harder surfaces.

The three most used cam and follower systems are **Radial**, **Offset Translating Roller Followers**, and **Swinging Roller Follower**.

11-2   Gears

Gears are classified by the position or location of the shafts they connect. Example: spur, pinion or helical gears are connected to shafts that are parallel to each other. The intersecting shafts are at 90°, and are usually connected by a bevel gear or angle gear. Shafts that are not parallel to each other and that do not intersect use a worm gear. In order to connect rotary motion into reciprocating, or back and forth motion, a rack and pinion would be used.

There are many types of gears such as spur, rack and pinion, ring, bevel, angle, miter, spiral, and worm. Each is designed for a specific purpose.

In this section, three basic types of gears, **Spur**, **Bevel**, and **Worm,** will be discussed, with some examples of where and how they are used. Gears are used to transmit motion and power at a **constant angular velocity**.

Spur Gear – The spur gear is a circular gear with teeth cut around its circumference. Two mating spur gears can transmit power from one shaft to another parallel shaft (**Figure 11-7**).

Figure 11-7 Spur Gear

NOTE: When the two meshing gears are unequal in diameter the smaller gear is called the **PINION** and the larger is called the **GEAR**.

Tooth Forms – The most common gear tooth form is the INVOLUTE tooth with a 14.5° pressure angle. This is the angular point of contact between the two gears.

Involute – The standard gear face that keeps the meshing gears in contact as the gear teeth are revolved past one another.

The **involute curve** can be thought of as the path of a string that is kept taut as it is unwound from the base arc.

Bevel Gears – Gears whose axis intersect at angles. Sometimes bevel gears are referred to as miter gears. This is the case when the gears are the same size, and the shafts are at right angles (think of them as rolling cones). The angles are usually 90° (see **Figure 11-8**).

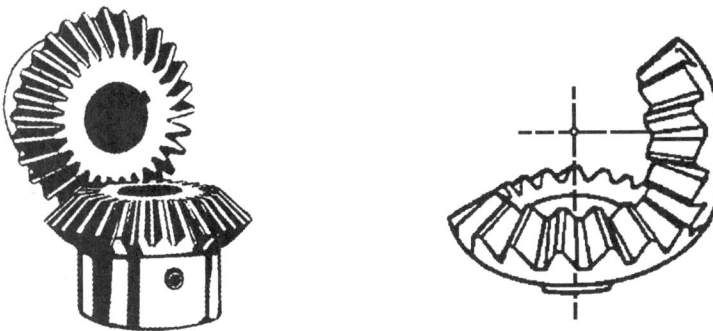

Figure 11-8 Bevel Gears

NOTE: The **smaller** of the two bevel gears is called the **PINION**, the same as in the spur gearing.

Worm Gears – Gears composed of a thread shaft called a **WORM** and a circular gear called a **SPIDER**. The worm is revolved in a continuous motion, causing the spider to revolve about its axis. The worm gear drives the spider gear. The worm is similar to a screw. It may have single or multiple threads. You would use this system to transmit motion between two perpendicular non-intersecting shafts. It is used primarily to change high speed to slow speeds (see **Figure 11-9**).

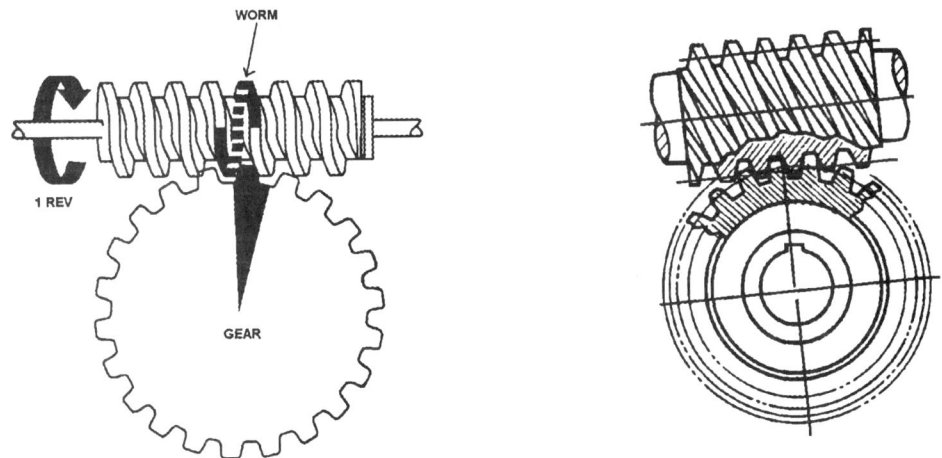

Figure 11-9 Worm Gears

11-3 Sprockets, Idler Sprockets, and Chain Drives

Sprockets – Sprockets are used to drive a chain, transmit torque, and increase or decrease RPM's. The basic sprocket when used with precision steel roller chains conform to ANSI (American National Standards Institute) standards (see **Figure 11-10**). They are used for mounting on flanges, hubs, or other devices. The plate sprocket is a flat, hubless sprocket. The standard provides for two classes of sprockets designated as Commercial and Precision.

Figure 11-10 Basic Sprocket with Chain

There are four basic types or designs of roller chain sprockets (see **Figure 11-11**).

1. plain plate
2. hub on one side only
3. hub on both sides
4. detachable hub

In addition, shear pin and slip clutch sprockets are designed to prevent damage to the drive or to other equipment caused by overloads or stalling.

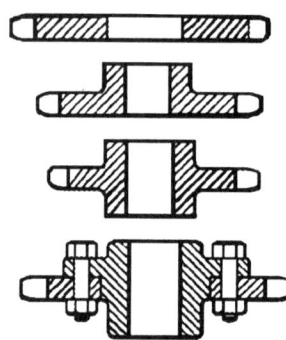

Figure 11-11 Sprocket Designs

NO ADJUSTMENT

Idler Sprockets – When sprockets have a fixed center distance or are non-adjustable (**Figure 11-12**), it may be advisable to use an idler sprocket to take up the slack.

Figure 11-12 Non-adjustable Sprocket

The idle is usually placed between the two chains, on the slack side of the chain (see **Figure 11-13**). This will allow the drive side of the chain to run straight.

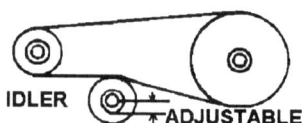

IDLER   ADJUSTABLE

Figure 11-13 Adjustable Idler Sprocket

Chain Drive – This is the transmission of power from an actuator to a remote mechanism by means of a flexible chain and mating toothed sprocket wheel.

Nearly all types of power-transmission chains have two basic components: side bars or link plates, and pin and bushing joints. The chain articulates at each joint to operate around a toothed sprocket (see **Figure 11-14**). The pitch of the chain is the distance between centers of the articulating joint.

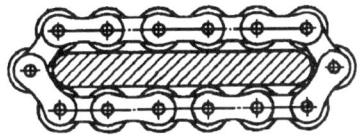

Figure 11-14 Drive Chain

They have several advantages: relatively unrestricted shaft center distances, compactness, ease of assembly, elasticity in tension with no slip or creep, and ability to operate in relatively high temperature.

There are many types of power-transmission chains with numerous modifications and special shapes for specific applications. The bead chain is often used for light-duty applications.

11-4   Pulleys, Belts, Idlers, Ratchet Gears, Bushings, Segment Gears

Pulleys and Belts – They are used for shock absorption, efficient power transmission at high speeds, flexibility, resistance to abrasive atmosphere, and are comparatively low in cost (see **Figure 11-15**).

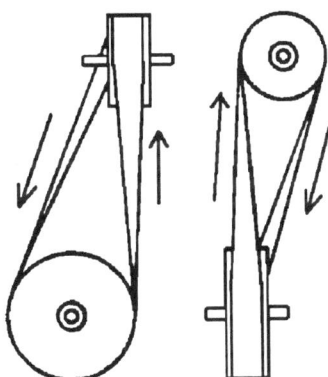

Figure 11-15 Pulleys & Belts

They transmit torque, and increase or decrease RPM's. They can operate on relatively small pulleys and be spliced or connected for endless operation. However, because they require high tension, they also impose high bearing loads. They are sometimes noisier than other drive belts. They can slip and have comparatively low efficiency at moderate speeds.

Idlers – Devices used between the driver and the driven unit to transfer motion. It can also be used to support or guide gears, chains, and/or belts (see **Figure 11-16**).

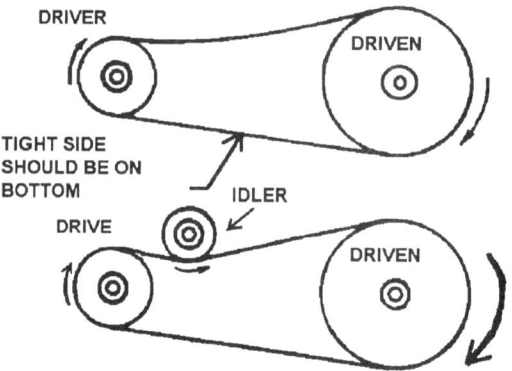

Figure 11-16 Idlers

Ratchet Gears – Ratchet gears are directional gears (**Figure 11-17**). Their teeth are cut on an angle that determines its direction. They are used on devices that require switches to rotate in one direction. Reversing the direction may cause damage to electrical contacts.

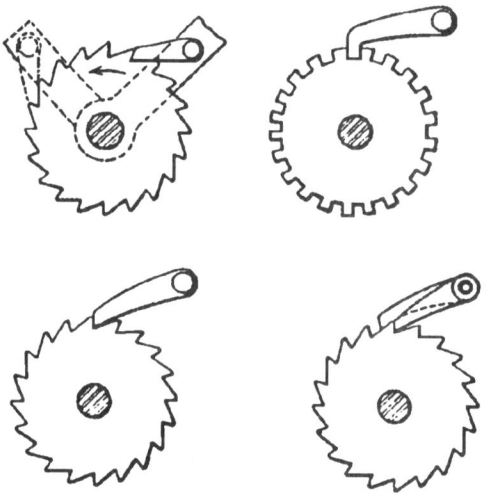

Figure 11-17 Ratchet Gears

They are also used when connected to mechanical counters. They may be used to transmit intermittent motion, or their only function may be to prevent the ratchet wheel from rotating backwards.

Pawl – A device used in conjunction with the ratchet gear to transfer mechanical motion to other devices used in the machine. It primarily controls the direction of the ratchet gear. It can be used to suspend loads. The pawls can be used with reversing ratchet gears used in reciprocating devices (either direction) (see **Figure 11-18**).

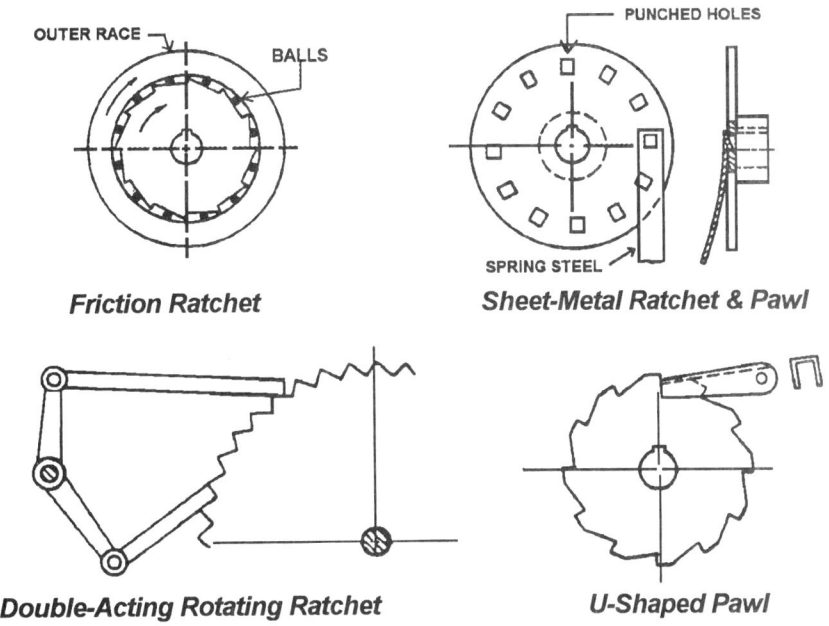

Friction Ratchet   *Sheet-Metal Ratchet & Pawl*

Double-Acting Rotating Ratchet   *U-Shaped Pawl*

Figure 11-18 Pawls

A **Segment Gear** (**Figure 11-19**) is a special type of gear with teeth on only one portion of its circumference. It is normally considered to be classified with the rack and pinion gears. Its motion is that of reciprocating (back and forth). This type of motion can permit the raising and lowering of mechanisms or to control the lifting mechanisms.

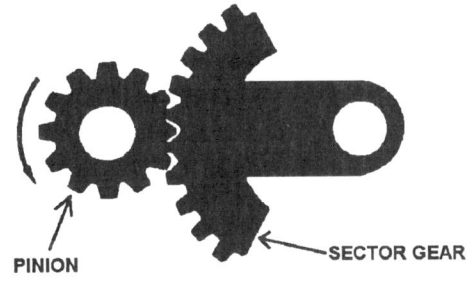

Figure 11-19 Segment Gear

Figure 11-20 is a practical application of a ratchet mechanism. It is a **jack ratchet** which uses two pawls to control the jack.

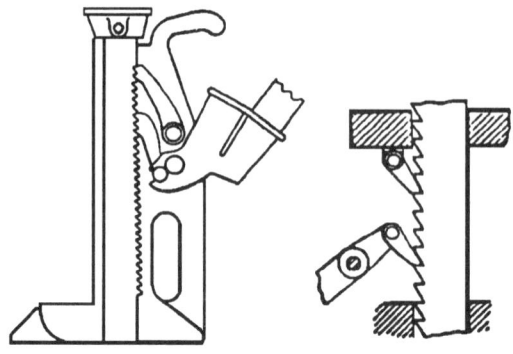

Figure 11-20 Ratchet Jack Mechanism

11-5   Timing Belts, Clutches, Thrust Bearings, and Bushings

Timing Belts (Figure 11-21) - Belts with grooves that run across the belt and mate with gear teeth are used for timing sequences. They are used on devices which require positive movement, similar to roller chains, with the difference being in the light weight of the belt vs. chains. In addition, timing belts drive devices that require indexing in degrees rather than continuous movement.

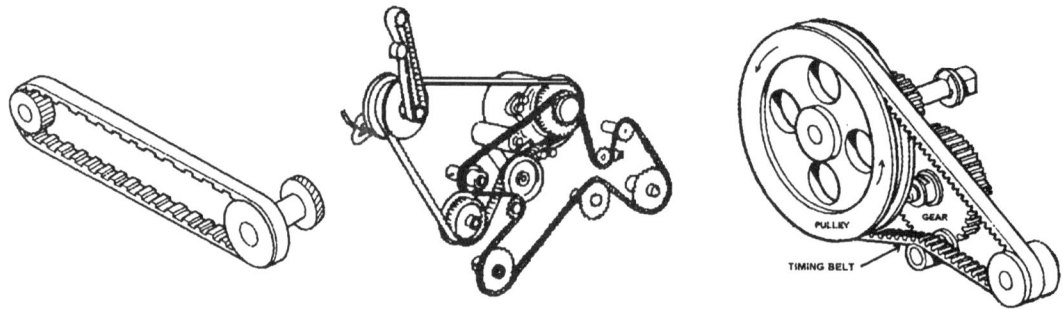

Figure 11-21 Timing Belts

Clutch (Figure 11-22) - A device that is used to relieve or exert torque pressure on a shaft or motor assembly. When the driving and the driven device members of a clutch are connected by the engagement of interlocking teeth or projected lugs, the clutch is said to be "positive", that distinguishes it from the type in which power is transmitted by frictional contact.

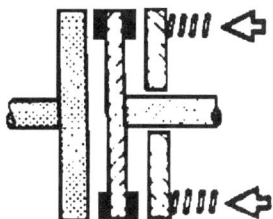

Figure 11-22 Friction Clutch

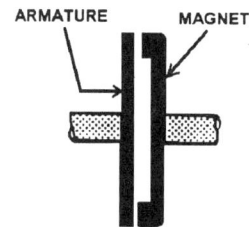

Figure 11-23 Electric Clutch

Frictional clutches can be either mechanically or electrically controlled, and can be used to change directions of the shaft. **Figure 11-23** is an electric clutch that uses electromagnetism to control its action. In **Figure 11-24**, there are four clutch couplings which show different types of engaging mechanisms, all performing the same function, to give a positive engagement. The positive clutch is used when sudden starting action is normal and when the inertia of the driven parts is relatively small. There are various designs of positive clutches that differ mainly in the angle or shape of the engaging surfaces.

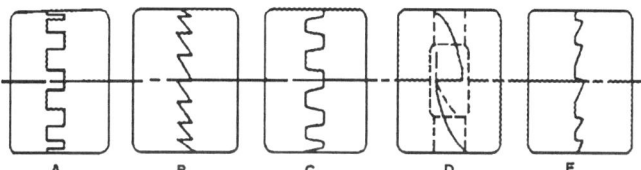

Figure 11-24 Clutch Couplings

Thrust Bearings - As the name implies, thrust bearings are used to either absorb axial shaft loads or to hold shafts position. In some applications they are used in place of ball-bearings due to their low torque requirements. There are many types of thrust bearings such as the Flat Plate, Step, Tapered Load, and the Tilting Pad. The most common type is the Flat Plate. It is the lowest in cost and the most frequently used. It is primarily used as a positioning device. **Figure 11-25** shows four types of thrust bearings highlighting the oil location and contact surfaces.

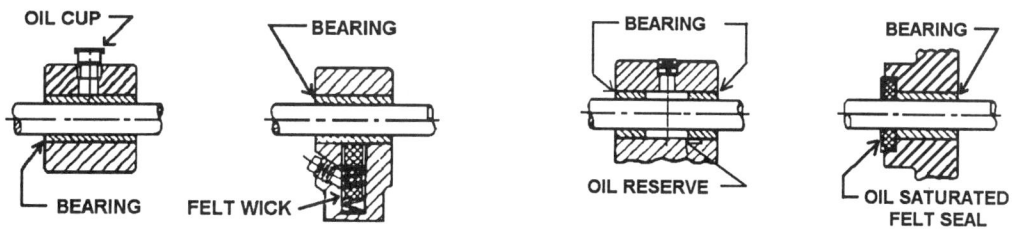

Figure 11-25 Thrust Bearings

Bushings – Bushings are used as hole reductions, spacers, standoffs, and slip bushings. They are also adapted for other uses such as wear sleeves and liners. They look like a short metal tube that is machined inside and out to precise dimensions, and are usually made to fit into a machined hole. They are usually made slightly larger to permit press fitting into holes. **Figure 11-26** shows a bushing that can be used in many different applications.

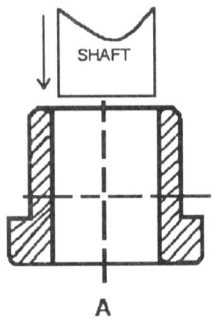

Figure 11-26 Bushing

Summary

This chapter covered many basic types of mechanical devices. Each type presented, from cams to bushings, is used in some mechanical device. The devices discussed range in all sizes and are made of different types of materials. Mechanical devices from cars to printers to disk drives and VCR's all require some type of mechanical device to move some thing or some one. As a technician, it is necessary to be able to use technical manuals to make repairs or adjustments. Therefore, a basic understanding of the most common type of mechanical device is essential to reading service manuals. It is needed to aid in repairs or adjusting of mechanical devices. In addition, identification and a basic knowledge of their function will aid in making successful repairs of mechanical devices.

SECTION III – MECHANICAL TECHNOLOGY

CHAPTER 12 – MECHANICAL RATIO

Introduction To determine the ratio of one pulley to another or to determine the size of an unknown pulley lies in a basic understanding of a simple algebra formula. It is called the inverse ratio formula. There are two basic words used in this formula, they are **DRIVER** and **DRIVEN**. If you can remember those two words, and apply those procedures, covered in this section, then you will become proficient at calculating any ratio.

Objectives After completing this chapter, you should be able to:

- ◆ Determine the RPM's of a driven pulley when the proper values of a driver pulley and the size of the driven pulley are known.

- ◆ Determine the size of a driver or driven pulley either RPM's and size of a driver pulley is known.

- ◆ Determine the direction of a driven set of pulleys when the direction of any one pulley is known.

- ◆ Determine a gear's ration when two sets of parameters on any gear and or the RPM's or number of teeth on any attached gear is known.

- ◆ Determine the amount of effort needed or distance to lift an object when the number of ropes or number of pulleys are known.

- ◆ Determine any speed of any belt driven device when the proper information is known.

12-1 Single Pulley Ratio

In each pulley ratio problem there must be a driver pulley and a driven pulley. The DRIVER is identified by the two sets of parameters; the RPM's and the SIZE of a pulley. Once that has been determined, the formula can be used. The way it is set up will be **of little importance** provided that you are consistent in your set up.

Since the formula is an equation, it means that an equality must be maintained. In an equation, whatever you do with one side of the equation, you must do to the other.

One side of the formula is reserved for the sizes of a set of pulleys and the other is reserved for the RPM's of the same pulleys.

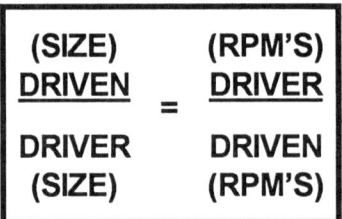

Figure 12-1 An Example of the Single Ratio Basic Formula

Note the inverse of the driver and driven words, they are the opposite of each other.

Example (see **Figure 12-2**)

Determine the RPM's of the DRIVEN pulley if the **DRIVER** is 10" and turning at 1000 RPM's and the **DRIVEN** pulley is 5" in diameter.

Placing the numbers on the top and bottom of the formula sets up the equation.

Since this is a ratio formula, the procedure is to **cross multiply** the numbers and the letters.   5 (x) X = 5X and 10 (x) 1000 = 10,000 therefore;

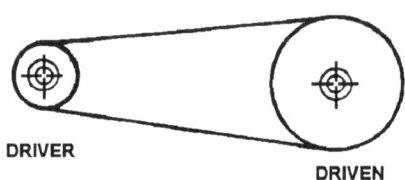

Figure 12-2 Single Pulley   *Setup Pulley Equation*

5x = 10,000, to determine the unknown divide both sides of the equation by 5. By doing this step you will determine (x) or obtain the unknown by itself that the driven pulley is turning at 2000 RPM's.

To solve for the size of a pulley, the procedure used in solving the RPM's of a specific pulley is applied. The only difference is that the unknown is the size, instead of the RPM's. Cross multiply and solve for the unknown. Note it's the same math procedure as in solving for the unknown RPM's.

12-2 Multiple Pulleys

To calculate multiple pulleys, that is, pulleys with more than one set of belts, the same basic formula is used as in the single pulley. Each set of pulleys is treated as a single set of pulleys, comprising a driver and a driven pulley. Solve for the first set of pulleys, then use the new information to apply to the second set of pulleys. (Keep in mind that pulleys on the same shaft rotate at the same RPM's.) The process is repeated until all unknowns have been solved.

Figure 12-3 is a set of multiple pulleys. Note the driver and the driven pulleys. They are connected by a belt, then note the second set of pulleys are also connected by a belt. To solve for the first unknown, apply the original basic formula, **DRIVER ÷ DRIVEN**, selecting the pulley with two sets of parameters as the driver. Then apply the basic formula **DRIVER ÷ DRIVEN** again to the second set of pulleys. Make the new driver the pulley containing the new set of parameters. Determine the first unknown, and then apply it to the second set of pulleys. Solve for the final unknown.

Figure 12-3 Multiple Pulleys

12-3 Gear Ratio

The gear is a toothed device used to transmit motion from one point to another. It is used to increase or decrease RPM's (rotations per minute) from one gear to another by means of the quantity of teeth on each gear and the RPM's of the driver gear. They are also used to change direction of rotating devices or driver gear.

To determine any of the unknowns, RPM's or number of TEETH, it is only necessary to remember the terms DRIVER and DRIVEN. The <u>driver</u> is normally identified by the two sets of parameters given to it; <u>size and number of teeth.</u> The formula used is an inverse ratio formula and it is also an equation. This means it is an equality, that whatever you do to one side of the equation you do to the other, to keep it a true equality (see **Figure 12-4**).

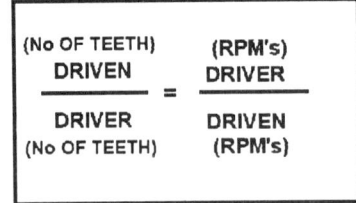

Figure 12-4 Gear Ratio

Example (see **Figure 12-5**): There are two gears, (A & B) "A" has 10 teeth and is rotating at 900 RPM's, "B" has 15 teeth but its speed is unknown. Determine how fast "B" is rotating.

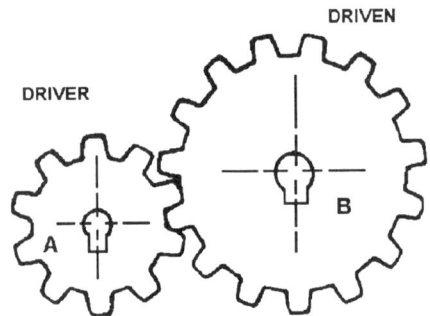

$$\frac{\text{(No OF TEETH) DRIVEN} \quad 15}{\text{DRIVER (No OF TEETH)} \quad 10} = \frac{\text{(RPM's) DRIVER} \quad 900}{\text{DRIVEN (RPM's)} \quad B}$$

Figure 12-7 Gear Example   *Setup Pulley Equation*

To solve for the unknown, insert the numbers in the formula and cross multiply.

15 (X) B = 15B   10 (X) 900 = 9,000

15B = 9,000

Divide both sides of the equation by 15 leaves B by itself on one side and 600 RPM's on the other.

Therefore "B" is traveling at 600 RPM's.

If the number of teeth are to be found, use the same formula but substitute the unknown number of teeth in the formula and solve as in RPM's. That is, cross multiply and solve.

12-4   Multiple Gears

The term multiple gears refers to more than two gears joined to perform a task (see **Figure 12-7**) The RPM's and the direction of the gears is related to the number of teeth and RPM's of the driver gear. To determine the RPM's of the driven gear apply the same formula as was used in single gear determination.

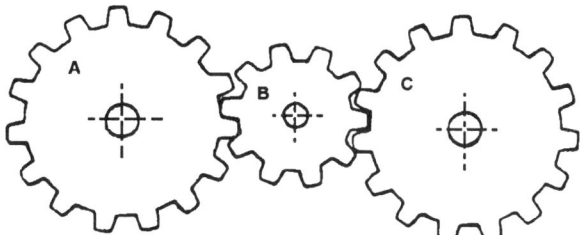

Figure 12-7 Multiple Gears

Once the RPM's of the driven gear has been determined, it then becomes the new driver for the next gear and so on until the final gears RPM's have been solved.

Here is an example of multiple gears calculations.

> **Note:** In the example, the illustration's quantity of teeth do not correspond to the numbers given in the written example.

Example: There are three gears connected together in a string (Gears "A", "B", and "C"). Gear "A" has 20 teeth and is rotating at 200 RPM's clockwise, Gear "B" has 10 teeth, and Gear "C" has 40 teeth. How fast is gear "C" rotating and what is the direction of the gear, is it rotating clockwise or counter clockwise?

Solution: Applying the basic formula. The DRIVEN "B" has 10 teeth but its RPM's are unknown. The DRIVER, "A" has 20 teeth and is rotating at 200 RPM's. How fast is the DRIVEN gear rotating?

$$\frac{10}{20} = \frac{200}{\text{"B"}} \quad \text{cross multiply the numbers and letters}$$

10 (x) B = 10B   and   20 (x) 200 = 4000

Therefore: **10B = 4000**, remember the rules of an equation. To maintain equality, divide both sides of the equation by 10. This leaves "B" by itself on one side. Repeating the same process to the other side of the equation leaves 400. This means that the driven gear is rotating at 400 RPM's.

To solve for gear "C" repeat the same process as you did for gear "B". This time make "B" the driver gear and gear "C" the driven, using the RPM's of gear "B".

$$\frac{40}{10} = \frac{400}{\text{"C"}} \quad \text{cross multiply the numbers and letters}$$

$$40 \times C = 40C \quad \text{and} \quad 10 \times 400 = 4000$$

Therefore: 40C = 4000, To solve for "C" divide both sides of the equation by 40. This leaves "C" on one side and 100 RPM's on the other. By performing this procedure on any multiple set of gears, the unknown RPM's can be found.

If the number of teeth are to be found, use the same formula, but substitute the unknown number of teeth in the formula and solve as in RPM's, that is, cross multiply and solve.

To determine the direction of the gears start with the gear that the direction has been given. Remembering that the teeth of each gear pushes the teeth of other gear causing it to turn in the opposite direction. Each gear pushes the other in the opposite direction. **Example:** There are three gears connected in series. The first one is set to turn cw (clockwise), it pushes the second gear ccw (counterclockwise) and it in turn pushes the next gear cw.

To solve any gear ratio it is only necessary to determine the driver then apply and substitute the known quantities in the formula and cross multiply.

12-5 Block And Tackle Ratio

Block and tackle pulleys are used to reduce lifting force. This is accomplished in a special configuration of the pulleys with only one rope wound so as to permit a single operator to lift heavy objects with little effort. The effort and the height the objects can be lifted will depend on several variables. They are the quantity of pulleys used and the distance between each of the pulleys.

To understand the relationship of pulleys to effort a discussion of three basic types: 1) single pulley, 2) two pulley, and 3) three pulley systems.

Single Pulley

The first is the single pulley system. This consists of a single pulley and a single rope (**Figure 12-8**). The pulley is attached to a fixed point. Since there is a single rope and a single pulley, the ratio is 1-to-1. There is no mechanical advantage to this system.

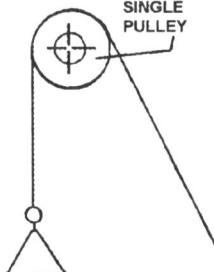

SINGLE PULLEY

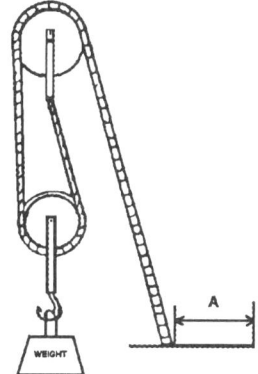

Figure 12-9 Two Pulleys

The second system consists of two pulleys with a single pull rope (**Figure 12-9**). This system of pulleys provides the operator the mechanical advantage of 2-to-1. The rating is obtained by taking the number of pulleys and dividing that number into the weight to be lifted. The height the object can be lifted is based upon the actual distance between the two pulleys.

A second calculation required on most pulley systems is the amount of rope that is to be taken up to lift the object. This calculation is based upon the lifted distance times two.

Example: A 200 lb. weight is to be lifted from a platform, and if the distance between the two pulleys prior to lifting the weight is 10". How much effort is needed and how much rope will be taken up to lift the weight?

There are two pulleys; therefore the effort is obtained by dividing the two pulleys into the weight, which makes the effort 100 lbs. This system is a 2-to-1 ratio.

The distance between the two pulleys is 10", therefore, multiply 2 times the 10", which makes the amount of rope to be taken up, 20".

Three Pulleys

The third system is a three-pulley system with a single rope. This system proves to be the most efficient. With three pulleys and one rope the effort is 3-to-1.

Example: If 300 lbs. is to be lifted 10" off the platform how much effort is needed, what must the distance be between the pulleys, and how much rope must be taken up to lift it (**Figure 12-10**).

Solution: If there are three pulleys the effort is found by dividing the three pulleys into the weight to be lifted. Since it is 300 lbs. and 3 pulleys, the effort is 100 lbs. which makes the effort 3-to-1. Distance between pulleys must be 10". Thirty inches of rope must be taken up to lift the 300 lbs. 10".

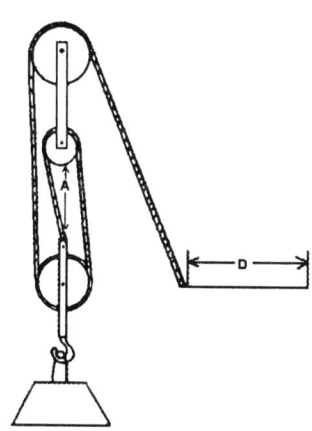

Figure 12-10 Three Pulleys

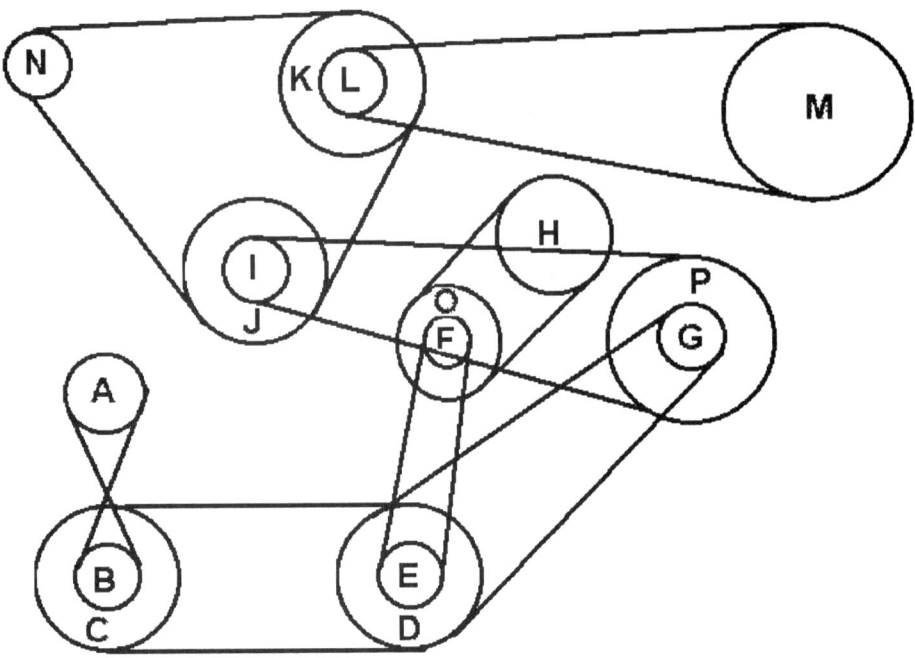

12-6   Pulley Ratio Worksheet

Pulley Ratio Exchange		
Pulley	**Size**	**RPM's**
A	6"	1725
B	3"	
C	8"	
D	12"	
E	14"	
F	2"	
G	3"	
H		3350
I	4"	
J	12"	
K	9"	
L	6"	
M		2150
N	14"	
O	8"	
P	9"	

12-7   Drill Press Worksheet

Determine the speed of the chuck on the drill press, if the belts are connected according to the chart below.

Determine The Speed of the Chuck	
	RPM's of Chuck
Belt is connected B to A	
Belt is connected D to C	
Belt is connected F to E	

Pulley	Size
A	5"
B	14"
C	8"
D	8"
E	14"
F	5"
G	10"
H	6"
I	15"
J	10"

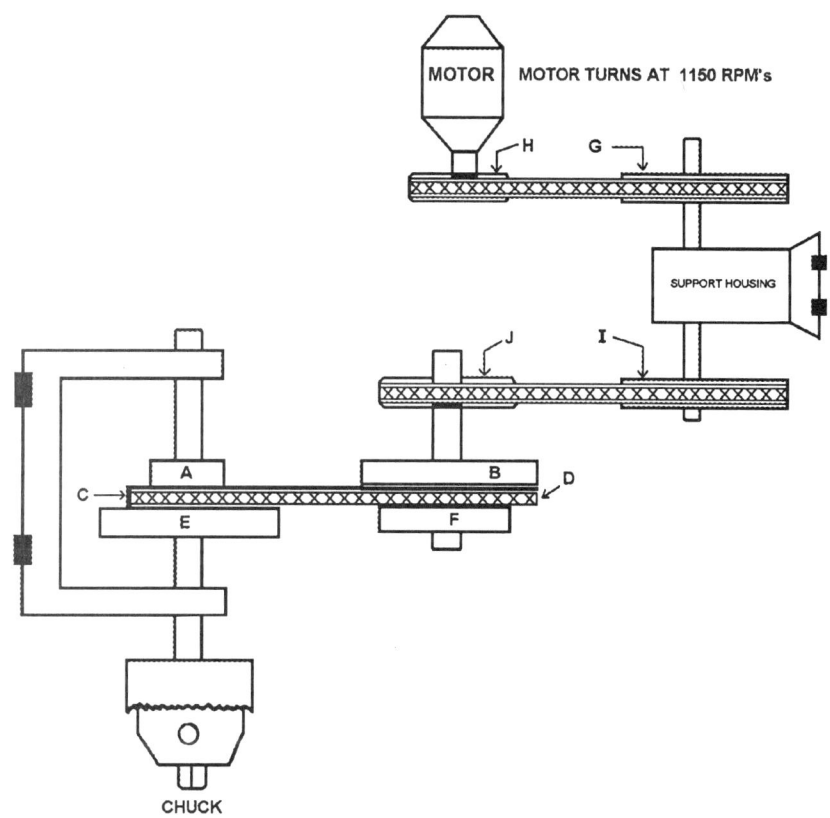

12-8   Pulley Worksheet

#	Sizes	RPM's
Pulleys		
1	16	
2	18	
3	21	
4	33	
5	42	
6	78	
7	12	

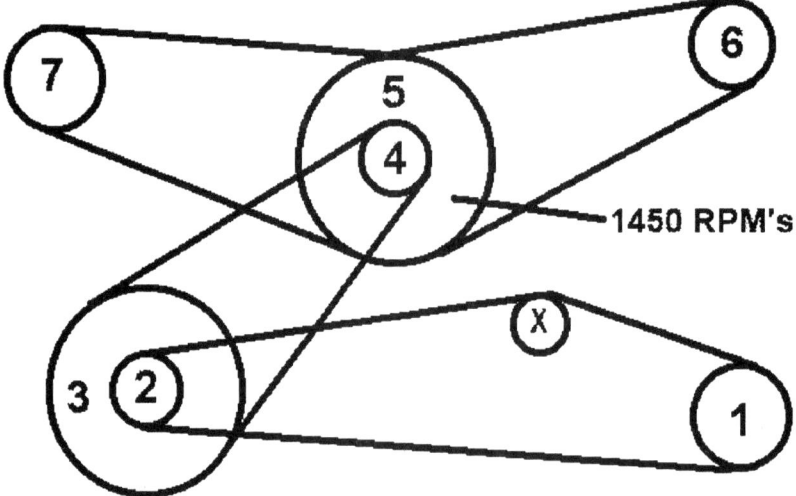

1450 RPM's

12-9   Gear Worksheet

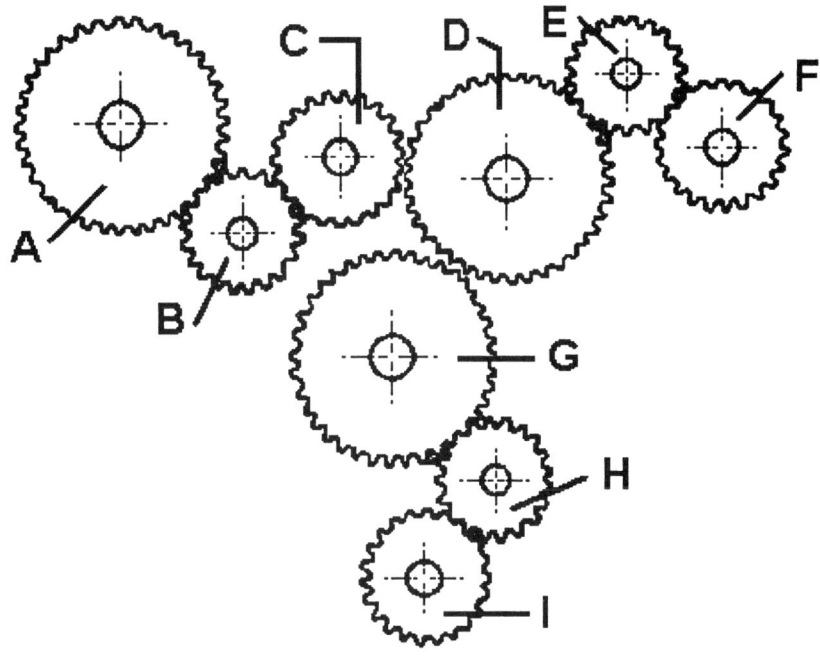

Gears			
#	# Teeth	RPM's	Direction
A	25 T		CW
B	32 T		
C	86 T	2800	
D	64 T		
E	14 T		
F	86 T		
G	32 T		
H	36 T		
I	42 T		

Summary

This Chapter presented the proper procedure for calculation of pulley, gear, and block and tackle pulleys. By understanding the formulas and practicing the problems, the technician will have a better understanding of how pulleys and gears are determined which may be helpful in understanding the operation of various mechanical devices. When making calculations, it should be noted that there are times when the diagrams shown and the numbers presented can be deceiving. Remember to use the math and not the picture sizes for proper determination. When diagrams are made, emphasis may not be on actual size relationships.

SECTION III – MECHANICAL TECHNOLOGY

CHAPTER 13 – ELECTRICAL TERMINOLOGY

Introduction   This Chapter covers basic electrical terminology as it relates to power for machines and electrical components. The technician should be aware of power sockets and their voltages. It also discusses the various forms used in reading electrical diagrams. It is important to know how power is supplied to wall sockets. When connecting to wall sockets, if the wiring is not correct damage to equipment may occur. In addition, there are other terms that can be useful in understanding basic electrical terminology.

Objectives   After completing this chapter, you should be able to:

- Draw and label a single cycle waveform.

- Draw and label a standard 120-volt wall socket.

- Draw and label a 220-volt wall socket.

- Identify and compare a schematic diagram, ladder diagram, and block diagram.

- Identify conduit and ground potential as they relate to electrical wiring.

13-1   Power Sources for Machines

120/240-volt Sockets

In most residential and business environments, three wires are used to bring electrical power from the outside into the house or business to a power distribution panel. In three-wire systems, the third wire is neutral and grounded in the distribution panel. The voltage between the other two wires is 240 VAC. Half of the voltage (120 VAC) appears between each of the other two wires and neutral. This 120/240-volt system is achieved by the power company at the last step-down transformer before it is distributed to the residence or business. The 240-volt secondary of the transformer is center-tapped, effectively splitting the 240 volts into two separate 120-volt lines in reference to a neutral. By connecting equipment between the two HOT wires, 240 volts is achieved.

120-volt Sockets

In systems of this type, the 120-volt load is divided as evenly as possible, within the distribution panel, and each circuit is supplied power through a circuit breaker of appropriate size to fit the load current and wire size used. A second wire for each circuit is connected to the ground bus within the distribution panel to serve as the neutral/return for each circuit. A third wire in each circuit is connected to the ground bus in the distribution panel and is routed with the hot and neutral to serve as a protective ground, intended to be physically attached to the housing, case, chassis, etc. of the attached equipment, when required. All current in the circuit must be carried by the HOT and neutral wires. The protective ground should never supply the return path except in the case of ground fault in equipment supplied by the circuit.

240-volt Sockets

Some heavy appliances require 240 volts. These 240-volt circuits consist of three wires; both HOT 120-volt lines and projective grounds. The loads are connected between the two 120-volt lines, yielding 240 volts across the loads. The third wire is normally connected to the housing, case, chassis, etc., of the attached equipment. The protective ground should never carry current except in case of a ground fault in the equipment supplied by the 240-volt circuit. Both sides of each 240-volt circuit are protected by a pair of "ganged" circuit breakers; therefore, a short or fault on one of the two HOT wires will cause both sides of the line to be opened by the ganged circuit breakers.

AC Cycle

One containing a nominal voltage of 120 VAC ± 5% on residential, and 120 VAC ± 7.5% on commercial. The socket contains three wires. The National Electric Code (NEC) specifies the wire color, wire voltage, amperage, and the location of the wires on the plug for both residential and commercial buildings.

Note: The voltage supplied is in the form of a single wave form. It is one complete revolution of a wave form starting at 0° rising to 90° for maximum positive voltage then down to 180°, then falling to maximum negative voltage in the opposite direction, at 270°, then back to 0° which is accomplished at 60 times per second (see **Figure 13-1**). The reason for identifying the sine wave form lies in the fact that the wave form peaks only twice in the cycle. It should be noted that when using a VOM meter to test a circuit the true voltage is read in RMS, (or effective) voltage, and not the peak voltage. The RMS or effective value is considered 70.7% of the peak value.

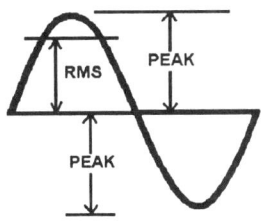

Figure 13-1 AC Waveform

STANDARD WALL SOCKET
(found in both residential, and commercial buildings)

The wire color and location on the electrical plug (see **Figure 13-2**) are listed below:

Wire Colors
1. HOT LEAD - black in color
2. NEUTRAL - white/gray in color
3. GROUND - green or bare copper or the conduit can provide ground.

Voltage measurements are made as follows:
1. Hot to Neutral - 120 VAC
2. Hot to Ground - 120 VAC
3. Neutral to Ground - 0 volts

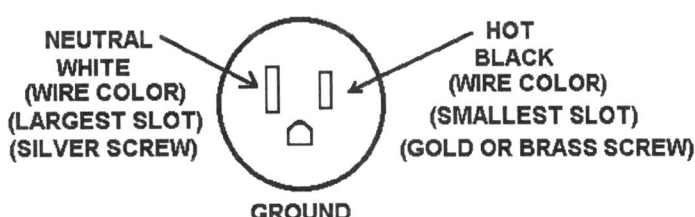

Figure 13-2 120 VAC Socket

Note: It should be noted that not all 240-volt sockets have the four connections as shown in **Figure 13-3**. Many have only three connections eliminating or configuring the socket without the neutral as shown in **Figure 13-4**. Therefore several different configurations of sockets are used, mainly to reduce the possibility of incorrectly connecting a piece of equipment.

<u>Wire Colors</u>
1. HOT LEAD - black
2. HOT LEAD - red
3. NEUTRAL - white/gray
4. GROUND - green or bare copper

<u>Voltage measurements are as follows</u>:
1. Hot to Hot 240 VAC
2. Either Hot to Neutral 120 VAC
3. Either Hot to Ground 120 VAC
4. Neutral to Ground "0" Volts

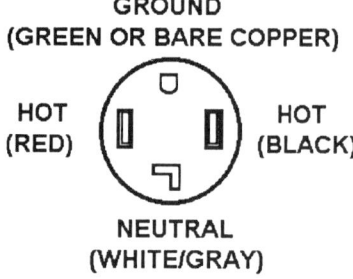

GROUND
(GREEN OR BARE COPPER)

HOT
(RED)

HOT
(BLACK)

NEUTRAL
(WHITE/GRAY)

Figure 13-3 240 VAC Socket

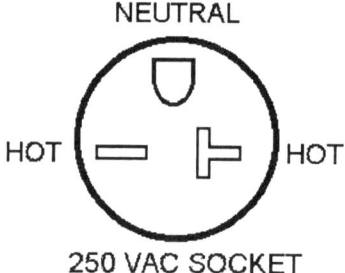

NEUTRAL

HOT

HOT

250 VAC SOCKET

Figure 13-4 240 VAC Socket

13-2   Industrial Electrical Terms

Schematic Diagram – This type of diagram shows all components of a circuit and their interconnections using standard symbols. Symbols on schematics may differ from those used in industrial machines. The diagram shows electrical circuits in the OFF position and all switches in their rest position or unactivated position. The advantage of this diagram is that the technician can energize the circuits that were presumed faulty. The disadvantage of this diagram is that the technician must read and determine which circuits are activated during the troubleshooting procedure. Schematic diagrams are normally associated with electronic equipment (see **Figure 13-5**).

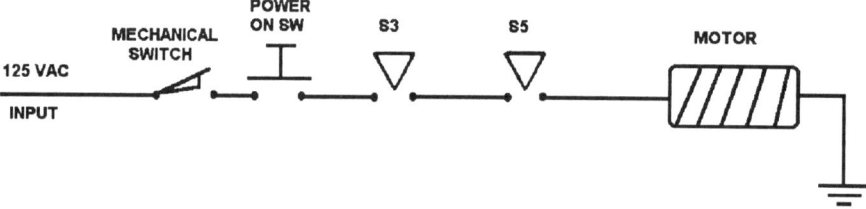

Figure 13-5 Schematic Diagram

Ladder Diagrams – This is a diagram that shows electrical circuits in a ladder form. It shows the inputs and outputs on either side of the rungs or power lines. All components are shown on horizontal lines (rungs of the ladder) (see **Figure 13-6**) which connect the two power lines. It can show electrical circuits in their energized condition or can show them in their de-energized position. The advantage of this type of diagram is that technicians can focus their time on the faulty section. The disadvantage of this type of diagram is that the problem may be originating from another section thus misleading the troubleshooter. This type of diagram is the standard way of representing circuits used to control industrial equipment and are normally associated with large machines rather than electronic equipment.

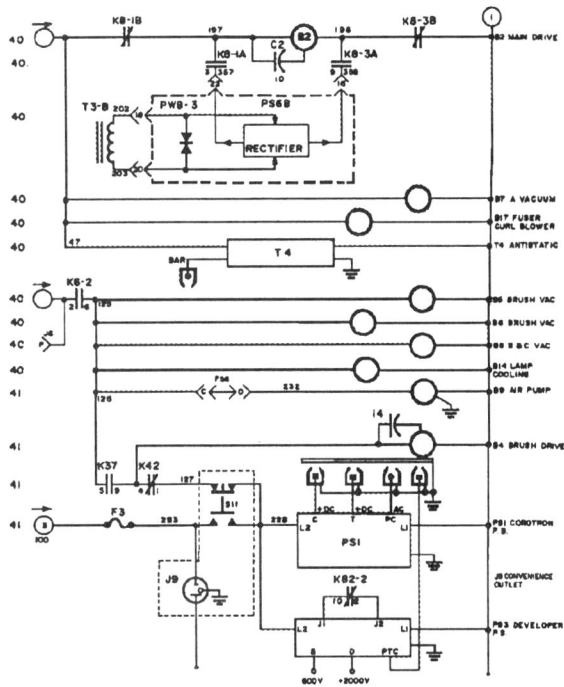

Figure 13-6 Ladder Schematic

Block Diagram – **Figure 13-7** shows a simplified schematic drawing. The block diagram shows direction through various subsystems. This enables the technician to check and recheck after repairs or adjustments have been made.

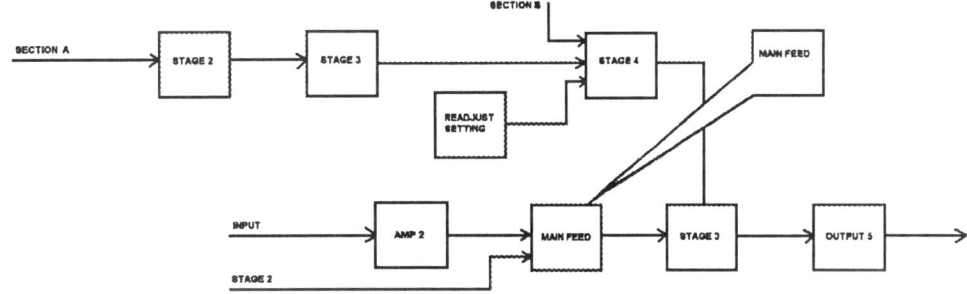

Figure 13-7 Block Diagram

Ground – A conducting connection, intentional or accidental, between an electric circuit or equipment chassis and the earth ground. The purpose of ground is to provide an alternate path for electricity to flow instead of having the operators provide that path. Grounds are used to prevent shock; therefore, care should be taken to ensure that the ground connection is solid.

Ground Potential – Zero voltage potential with respect to earth ground. This means that when checking ground with a meter, be sure that there is "0" voltage with respect to ground before working on any electrical outlet. In some cases a small amount of AC voltage may be present. This could be due to induction from other wires. If this occurs, verify that it is caused by induction and not by a faulty ground.

Conduit – Conduit is solid or flexible metal or plastic tubing through which insulated electrical wires are run. The purpose of the conduit is to prevent damage to the electrical wires. Damage can be from rodents chewing the insulation or vibration of machinery, which can also remove insulation from the wires, thus causing sparks which can cause fire. Metal conduit can act as a ground if the insulation melts.

Calibration – The determination of the deviation from a standard so as to ascertain the proper corrections. The definition refers to checking equipment for proper voltage settings and or proper dimensions so that they can operate safely or correctly. Calibration standards are maintained by the U.S. Bureau of Standards in Washington, DC. They provide industry with the standards with which to set their equipment. Without the Bureau of Standards, manufacturers would make their own standards.

Summary

Power for machines and reading wall sockets is essential to the technician. Schematic drawings are multi purpose. Check with manufacturer concerning their type of drawing. Grounds are important and should be checked. If improperly connected, they could cause injury to operators of the equipment and could result in extensive damage to the equipment. Calibration of all equipment is essential, from meters used to check equipment to the other devices that may be used in place of meters. The most essential part of checking for properly wired sockets is to verify before connecting equipment. Remember safety when checking voltages.

SECTION III – MECHANICAL TECHNOLOGY

CHAPTER 14 – MECHANICAL SWITCHES & CONDITIONS

Introduction   A mechanical switch is a device that acts as a controller for electrical circuits. In electronics it refers to the controlling of electrons. It does this by breaking the path or it can be used to complete a path, or divert electrons from one circuit to another. Switches can be used to momentarily start devices, as well as preventing start up of equipment. Switches are taken for granted until they malfunction or fail. It is important for the technician to know the capabilities and the characteristics of basic switches, how they are used in circuits, and their makeup.

There are thousands of switches on the market ranging in size, shape, current, and voltage capabilities. Many are designed for special purposes, others are made for general purpose applications, which is to conveniently turn on and off circuits. The classification of a switch is determined by its application. In some cases their name specifies where and how they are used.

This Chapter will cover many different types of mechanical switches including the conditions by which they function and, where appropriate, their schematic symbol.

Objectives   After completing this chapter, you should be able to:

- ◆ Identify basic switch functions.

- ◆ Draw and label the schematic symbol for a single-pole single-throw switch.

- ◆ Draw and label a self-latching relay circuit using a mechanical relay.

- ◆ Determine the purpose of a micro switch, or solenoid switch, and how it is used in a circuit.

- ◆ Determine the application of an interlock switch and draw and label a normally closed or a normally open switch.

- ◆ Identify the capabilities of a rotary switch.

- ◆ Identify poles and throws as related to switches.

- ◆ Determine the application of pressure switches, float switches, and circuit breakers.

14-1   Mechanical Switches

Switches are labeled on their terminal sides (the point to which the wires are connected). The points represent the function of the switch's internal contacts. **Figure 14-1** is an example of a **basic switch (a)** and its **schematic symbol (b)**.

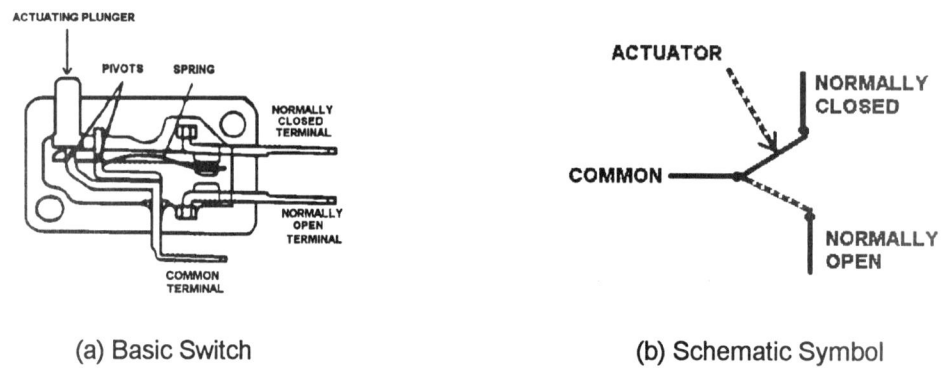

(a) Basic Switch   (b) Schematic Symbol

Figure 14-1 Snap Action Switch

14-2   Switch Functions

A switch is a controller for electricity in electrical devices. It prevents the flow of electricity or starts the flow of electricity by providing a path or acts as an interrupter of that path. Think of it as a draw bridge. Bridge down, traffic flows; bridge up, traffic does not flow.

Switches are used to prevent unauthorized entry into a machine by acting as a power interrupter. The example would be as office copier or a microwave oven. Each, if operated with the doors open, could harm the operator or damage the machine.

Switches come in all shapes and sizes, and they all function in a similar manner. They have movable contacts with an input and output connection. Some have multiple outputs and inputs, but all are mechanical in their internal movements.

In switches that have multiple movements, there are functions that are important to know. The term 'com' is the abbreviation for common. The other parts of switches with multiple functions are N.C. and N.O., which stands for normally closed and normally open. Both terms refer to a condition of the switch without any contact with the switch: "Out of the box".

COM – A part of the mechanical switch which is used by the normally open and the normally closed portion of the switch. It is common to both parts of the output terminals. **Figure 14-2** shows a schematic diagram for a two-position switch. Note that the com can be moved from one point to another and provide an electrical path for the current to flow.

Figure 14-2 Two Position Switch Schematic

Normally Closed (N.C.) – That portion of the mechanical switch that does not require the switch to be actuated. It will show continuity (allow a flow of current flow) in the rest position. **Figure 14-3** shows the normal connection of the switch at its rest position.

Figure 14-3 Switch at Rest

Normally Open (N.O.) – The part of a mechanical switch which requires that the mechanical switch be actuated by a mechanical device. When actuated, the switch will allow current to flow. **Figure 14-4** shows the switch moved from the **N.C.** to the **N.O.** position. The solid line indicates the new connection and the dotted line shows where it was moved from.

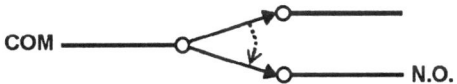

Figure 14-4 Normally Open Position

Voltage Rating – The voltage rating is the maximum voltage that the switch is designated to control. A voltage higher than the voltage rating of the switch may allow the voltage to "jump" the open contacts of the switch. This "jump" of voltage may cause the circuit to be on all the time with no way to turn it off. **Figure 14-5** shows a schematic symbol of a toggle switch. The distance between the contacts determines the voltage rating of the switch. In most cases the voltage rating of the switch is written on the switch.

Figure 14-5 Toggle Switch Schematic

14-3   Specialty Switches

Relay – A relay is called an electromechanical switch. It is used to switch on high current devices. The coil is an electromagnet. The coil requires only small amounts of current to operate it, but the electrical contacts that perform the switching **(on or off)** are able to take high (current) amperage. This device, when properly wired in a high current circuit, allows maximum safety for the operator. **Figure 14-6** is a self-latching relay circuit. It can have multiple sets of contacts which are used to control high voltage circuits.

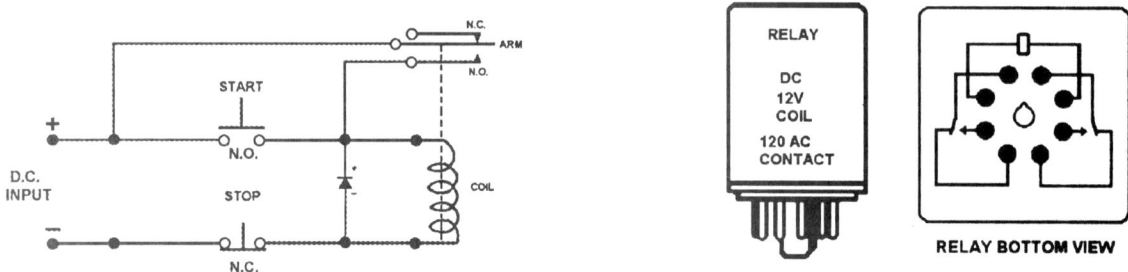

Figure 14-6 Universal Self-Latching Relay Circuit including Relay

Solenoid – The solenoid is called an electromechanical actuator. It actuates devices (starts up a device) or it can be used to open valves. It has an electromagnetic core. When it is energized with electricity, the movable core, or plunger, can move a small mechanical part a short distance. When it is turned off, a spring connected to the plunger is used to return it to the rest position. These devices are used on various machines from copiers to computers. They are also used extensively in consumer products and industrial machines. **Figure 14-7** shows a solenoid connected to a flow valve. In this case, it is used to open or close the valve.

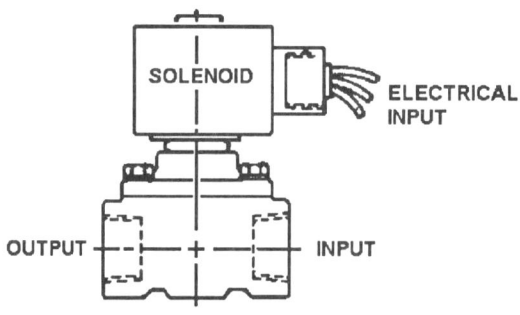

Figure 14-7 Solenoid Connected to a Flow Valve

Micro Switch – The micro switch (see **Figure 14-8**) is a precision snap-acting switch. The operating point is pre-set and very accurately known. This type of switch is used in parts of machines or devices that require precision timing. The switch's internal mechanism is designed to actuate at exactly the same time over and over. Other switches which have similar internal mechanical parts do not function as accurately. The micro switch plunger mechanism requires only light pressure to actuate it, whereas other similar type switches require greater pressure to perform the same function.

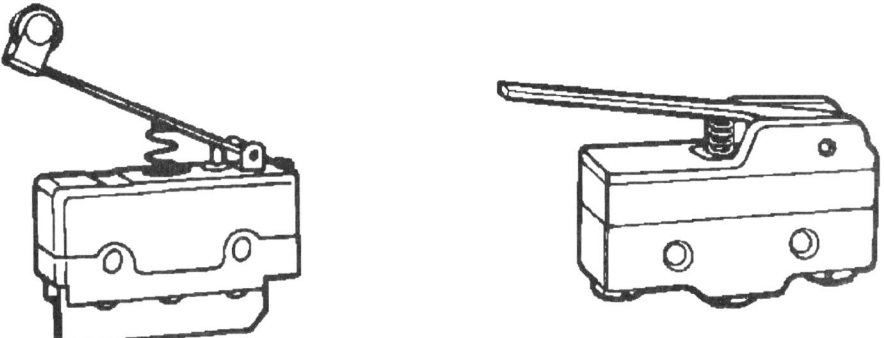

Figure 14-8 Micro Switches

The micro switch can be used to sense paper in printers or notify the operator that the printer is out of paper. They are used on household appliances to prevent the operator from becoming injured by interrupting the power when the door is open. They have a multitude of applications as sensing devices. It is a very useful switch.

Snap Action Micro Switches – These switches are used in areas where other mechanical devices actuate them. The toggle and the push button are manually actuated. The micro's are actuated by other devices. Their physical size can be as small as an eraser on a pencil or up to a one square inch. Sizes can vary as needed and application dictates. The sizes given are not absolute but nominal sizes. The current capabilities are normally very low, and they are used in conjunction with other devices such as relays to actually perform the switching of circuits. **Figure 14-8** shows different types of the snap acting switches.

Interlock – This switch is used to interrupt power and can be used as a safety switch which prevents unauthorized service on machines. This switch acts like a micro switch in interrupting power when the device is improperly operated. Unlike the micro switch, the interlock is not a precision switch and its internal mechanism is similar but not exact. This type of switch can be found on many appliances, office machines, or any other device to prevent access by unauthorized personnel. It can be used to turn on and off lamps as a convenience factor. The interlock is found on so many devices that they are too numerous to mention.

The interlock schematic symbols (**Figure's 14-9** and **14-10**) are a triangle between two lines; not touching or touching depending on whether it is a N.O. or N.C. switch.

Figure 14-9 N.O. Switch   *Figure 14-10 N.C. Switch*

14-4   General Purpose Switches

Toggle Switch – This switch is probably the most commonly used switch. This type of switch is very versatile, in that it can have single and multiple actuation (turn on or off many circuits at one time) performed at a single throw of the toggle. The number and quantity of contacts is determined by the application. They can be used for high current circuits. Toggle switches or maintained switches are sometimes referred to as on-off switches. The term toggle means to move between two points. **Figures 14-11** is a toggle switch, **14-12** shows the internal workings of a toggle switch, and **Figure 14-13** is a schematic of a toggle switch.

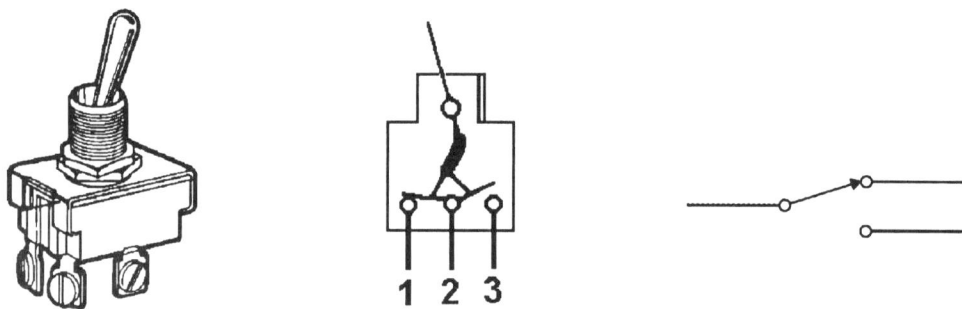

Figure 14-11 Toggle Switch   *Figure 14-12 Internal*   *Figure 14-13 Schematic*

Push Button Switch – The basic push button switch is used to momentarily actuate or deactuate an electrical circuit. This type of switch, like the toggle, can have single or multiple contacts based upon the user application.

Push button switches come in many shapes and sizes depending on their application. They can be push on - push off, momentary on, and can have as many poles as the circuit requires. They can be operated by a person or can be activated by a mechanical device.

Push button switches, just as the name implies, must be pushed to be activated either by an operator or an actuator. **Figures 14-14** and **14-15** show the schematic symbols for a momentary N.O. switch and a N.C. switch. **Figure 14-16** is a picture of a push button switch which is representative of either a N.O. or N.C. switch depending on internal design. The internal workings of push button switches are on each side of **Figure 14 -16**.

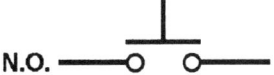

N.O.

Figure 14-14 N.O. Schematic Symbol

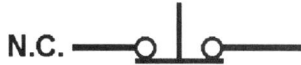

N.C.

Figure 14-15 N.C. Schematic Symbol

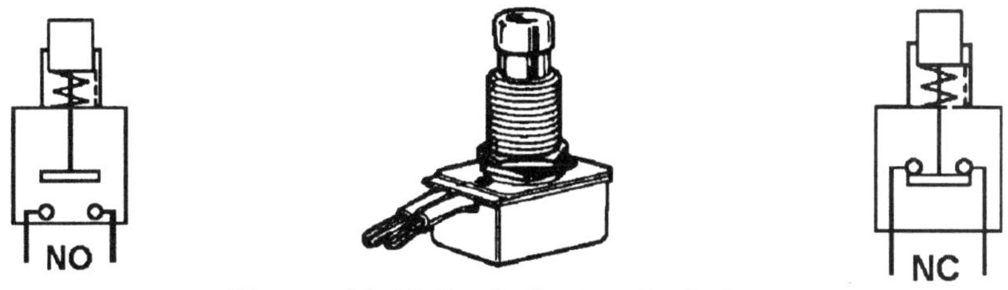

NO

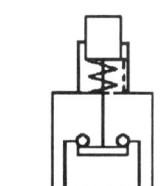

NC

Figure 14-16 Push Button Switch

Rotary Switches – This type of switch has the capability of switching many circuits simultaneously. Its construction is that of layers and each layer has multi-contacts that can be switched at one time. This type of switch is a low current switch. The switch can be constructed as circuit switching requirements dictate.

Rotary switches, as the name implies, rotate to make a selection. The application of rotary switches is extensive. They are used in electronic equipment or instrumentation to switch many circuits at once. One can be used in electronic receivers with many stations to select from. **Figure 14-17** shows a rotary switch in the top, side, and bottom views.

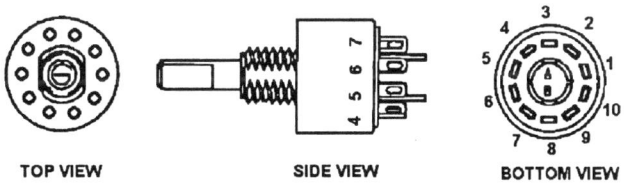

TOP VIEW   SIDE VIEW   BOTTOM VIEW

Figure 14-17 Rotary Switch

14-5   Additional Types of Switches

Switches are designated by the pole method: single pole-double throw, double pole-double throw, etc.

Poles refer to the movable arm in a switch that moves from one set of contacts to another which is used to connect or disconnect different circuits. **Figure 14-17** shows a rotary switch. This type of electro-mechanical switch has a rotary shaft connected to single or double terminals that can make or break many different circuits at the same time. Toggle switches have a limited number of poles, and are used to connect and disconnect circuits.

Throw – This refers to the number of contacts the switch is designed to function in the circuit. An example is a **SPDT**. This switch has a single pole, and is a double throw. Another example can be an **ON-ON** switch or an **ON - OFF - ON**. This indicates the position of the lever.

SPDT – This is called a Single-Pole Double-Throw switch. It is a switch that can have only two positions **on-none-on**, or **on-none-off**. The **none** between the two functions means the switch does not have a center position. Switches are designated by the number of positions that can be switched.

A single-pole double-throw switch can also have a center off, but it still only has two positions. Switches can also have more than a one pole. It can be a **DPDT** (double – pole, double-throw), **3PDT** (three-pole double-throw) and so forth as many as is needed in a given circuit and is available by the manufactures.

SPST (single pole single throw) – This switch has only one position: on-off. The letter designation denotes the condition. This type of switch can have more than one designation and still be a single throw switch. It can have as many poles as is needed for the circuit. Other designations: **DPST, 3PST, 4PST**.

DPDT (double pole double throw) – This switch functions the same as the **SPDT** by having two positions. It has double poles and can have an off position in the center. **Figure 14-18** shows the internal parts of a toggle switch having three positions.

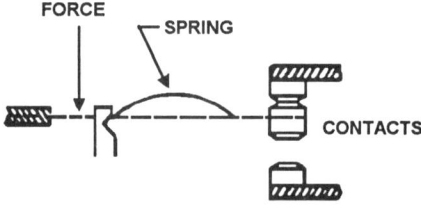

Figure 14-18 Pressure Switch

Pressure Switch – A pressure switch is defined as a switch that senses a change in force per unit area. A force cannot be seen but can be measured. The force or change in pressure can be detected by a sensor that is designed to move with any change. There are many different types of pressure switches available. They are used to sense air pressure change, liquid level change, dry power change, and even a change in vacuum.

The pressure switch is still a mechanical device. The mechanical parts are connected to a set of electrical contacts which form a switch. It is normally a simple on-off switch. The mechanism is comprised of the same materials found in micro switches which is a piece of beryllium copper connected to a set of silver contacts that transfer from one stationary set to another (**Figure 14-18**). When the contacts move, they usually activate a relay, which sends signals to the operator that some change has occurred.

As stated earlier, the pressure switch is used extensively in industry for determining levels of materials. They can be used in areas where it is hazardous to humans. They still use a basic switch for the transfer of movements.

Float Switch – The float switch is designed to measure liquid or dry materials in a closed container. The actual switch can be a simple REED switch with a set of magnetic contacts making a connection to a relay or it can be an RF signal with a phase shift sensor on the end of a probe. The phase shift occurs when materials come in contact with a probe. The phase shift then sends a signal to a relay which in turn notifies the operator that material is present.

The float switch utilizes the mechanical snap acting switches mechanism to activate the electrical circuits. The make up of the switch is the same as the pressure switch. The basic design of the micro switch is utilized in many different applications. A few of them were just covered, such as pressure switches and float switches.

Circuit Breaker – This is a switch that controls the flow of current by opening the circuit automatically when more electricity flows through the circuit than the circuit is capable of carrying. Resetting may be done either automatically or manually.

A circuit breaker is defined by the NEC as a device that can open and close a circuit by NON-AUTOMATIC means, and which opens the circuit automatically on a predetermined overload of current without injury to itself (when properly applied within its rating).

Most circuit breakers installed in residences contain thermal bi-metal elements for slow overloads and a magnetic arrangement for instantaneous trips in case of short circuits.

The circuit breaker looks like a toggle switch. Since they are used only on the hot side of the line, only one set of contacts is used for 120-volt branches. **Figure 14-19** shows a circuit breaker with its various positions. To restore energy from a tripped overload, the handle must be moved to "RESET" then to "ON".

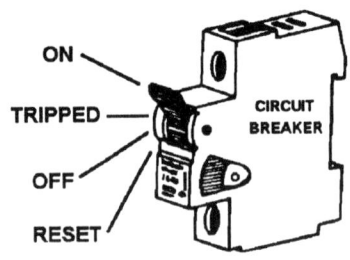

Figure 14-19 Circuit Breaker

Summary

Switches come in all sizes and designs depending on the needs for electronic equipment and industrial machines. There are many requirements for switches and many special applications. A few of the switches have been discussed. Conditions referred to how the switch is to be used in a circuit. Their letter designation refers to how they are to work or function in a circuit. Their voltage and current (AMPS) rating is designated by the manufacturer and is selected by circuit application.

WORK SHEETS ANSWER KEY

Chapter 6 Measuring Devices
Exercise 1 - Figures 1 to 12

1. 0.589
2. 0.022
3. 0.736
4. 0.083
5. 0.808
6. 0.388
7. 0.737
8. 0.598
9. 0.157
10. 0.679
11. 0.949
12. 0.520

Exercise 2 – Micrometer Reading

BARREL SCALE SETTING IS:	THIMBLE SCALE SETTING IS:	MICROMETER READING IS:
0.425" - 0.450"	0.016"	0.441"
0.075" - 0.100"	0.007"	**0.082"**
0.150" - 0.175"	0.003"	**0.157"**
0.875" - 0.900"	0.012"	**0.887"**
0.300" - 0.325"	0.024"	**0.324"**
0.000" - 0.025"	0.021"	**0.021"**
0.025" - 0.050"	0.013"	**0.038"**
0.750" - 0.775"	0.017"	**0.767"**

Chapter 7 Drilling
Drill Selection Exercise

Note: If answers match the selected answers, then drill selections were made correctly. Also note that when selecting a body size drill bit, select one 1 or 2 thousands larger to be sure of proper clearance.

Screw Size	Tap Drill Decimal	Tap Drill Number	Body Drill Decimal	Body Drill No.
0-80	3/64	0.047	0.060	53
1-64	53	0.060	0.073	49
3-48	47	0.079	0.099	39
5-40	38	0.102	0.125	1/8
8-32	29	0.136	0.164	19
10-32	21	0.159	0.190	11
1/4-20	7	0.201	0.250	1/4

Chapter 12 Mechanical Ratio
12-6 Pulley Ratio Worksheet

Pulley Ratio Exchange		
Pulley	Size	RPM's
A	6"	1,725
B	3"	3,450
C	8"	3,450
D	12"	2,300
E	14"	2,300
F	2"	16,100
G	3"	9,200
H	38.4	3,350
I	4"	20,700
J	12"	20,700
K	9"	27,600
L	6"	27,600
M	77	2,150
N	14"	17,742
O	8"	16,100
P	9"	9,200

Note: Pulleys on the same shaft rotate at the same RPM's G and I are on the same shaft therefore rotate at the same RPM's.

12-7 Drill Press Worksheet

Determine The Speed of the Chuck	
	RPM's of Chuck
Belt is connected B to A	2,898
Belt is connected D to C	1,035
Belt is connected F to E	369.6

12-8 Pulley Worksheet

Pulleys		
#	Size	RPM's
1	16	2,563
2	18	2,278
3	21	2,278
4	33	1,450
5	42	1,450
6	78	780
7	12	5,075

Note: Pulleys on the same shaft rotate at the same RPM's Both pulley 4 and 5 rotate at the same RPM's as do 2 and 3.

Gears			
#	# Teeth	RPM's	Direction
A	25	963.2	CW
B	32	752.5	CCW
C	86	280	CW
D	64	376.2	CCW
E	14	1,720	CW
F	86	280	CCW
G	32	752.5	CW
H	36	668.8	CCW
I	42	573.3	CW

INDEX